Sustainable Production, Life Cycle Engineering and Management

Series Editors

Christoph Herrmann, Braunschweig, Germany

Sami Kara, Sydney, Australia

SPLCEM publishes authored conference proceedings, contributed volumes and authored monographs that present cutting-edge research information as well as new perspectives on classical fields, while maintaining Springer's high standards of excellence, the content is peer reviewed. This series focuses on the issues and latest developments towards sustainability in production based on life cycle thinking. Modern production enables a high standard of living worldwide through products and services. Global responsibility requires a comprehensive integration of sustainable development fostered by new paradigms, innovative technologies, methods and tools as well as business models. Minimizing material and energy usage, adapting material and energy flows to better fit natural process capacities, and changing consumption behaviour are important aspects of future production. A life cycle perspective and an integrated economic, ecological and social evaluation are essential requirements in management and engineering.

Indexed in Scopus

To submit a proposal or request further information, please use the PDF Proposal Form or contact directly: Petra Jantzen, Applied Sciences Editorial, email: petra.jantzen@springer.com

More information about this series at http://www.springer.com/series/10615

Stefan Alexander Blume

Resource Efficiency in Manufacturing Value Chains

Stefan Alexander Blume
Institute of Machine Tools and Production
Technology
Technische Universität Braunschweig
Braunschweig, Germany

ISSN 2194-0541　　　　　　ISSN 2194-055X　(electronic)
Sustainable Production, Life Cycle Engineering and Management
ISBN 978-3-030-51896-7　　　　ISBN 978-3-030-51894-3　(eBook)
https://doi.org/10.1007/978-3-030-51894-3

This Springer imprint is published by the registered company Springer Nature Switzerland AG
The registered company address is: Gewerbestrasse 11, 6330 Cham, Switzerland

Supervisors' Foreword

The protection of our environment may be the greatest challenge of the twenty-first century. It demands a more responsible use of natural resources and a decoupling of economic growth from resource utilization. Since production makes a significant contribution to global resource consumption, the United Nations have identified the establishment of sustainable consumption and production patterns as one of 17 Sustainable Development Goals. Due to the organization of today's production, various cause-and-effect relations need to be considered when redesigning value chains to avoid problem shifting. However, the economic and environmental effects of local decisions on upstream and downstream processes of the value chain are usually not adequately considered in industrial practice today. A higher degree of transparency along value chains is a precondition to achieve significant progress toward ambitious sustainability targets. Therefore, decision-makers in the industry must be supported by adequate methods and tools to simultaneously and systematically pursue technical, economic and environmental targets.

With his work, Stefan Blume makes a valuable contribution to designing more resource-efficient value chains. He presents a concept for method-based decision support combining established and complementary methods such as material and energy flow analysis (MEFA), value stream mapping (VSM), life cycle costing (LCC) and environmental life cycle assessment (LCA). It enables holistic modeling and multi-dimensional performance assessment of manufacturing systems on different levels—from processes over value chains up to entire product life cycles. Systematic identification of hotspots, i.e., main levers to improve resource efficiency, is achieved by automated sensitivity analyses and benchmarking against reference processes. A knowledge-based system (KBS) supports in the selection of suitable situation-specific improvement measures. To facilitate the concept application, he developed a *decision support toolbox* as a user-friendly stand-alone software tool. The application of the concept is demonstrated using two case studies from the metal manufacturing industry. The results demonstrate significant resource-saving potentials of up to 30% savings in production costs and 40% savings in CO_{2eq} emissions. Altogether, the presented concept provides progress beyond the state-of-the-art regarding modeling, evaluation, and improvement of manufacturing value chains. The implementation as a decision support toolbox can foster continuous improvement processes in value chains in terms of technical, economic and environmental targets.

Prof. Dr.-Ing Christoph Herrmann
Technische Universität Braunschweig
Braunschweig, Germany

Prof. Dr. Sami Kara
University of New South Wales
Sydney, Australia

Acknowledgements

The present book was written in the context of my work at the Chair of Sustainable Manufacturing and Life Cycle Engineering of the Institute of Machine Tools and Production Technology (IWF) at Technische Universität Braunschweig. Herewith, I would like to express my sincere thanks to all those who have contributed to the creation of this book.

First of all, I would like to say a special thanks to Prof. Dr.-Ing. Christoph Herrmann as head of the institute. Thank you for always being a critical but constructive supporter. I very much appreciate your passion for research in sustainability, your constant will to make good things better, and your unshakable optimism. Further, I would like to thank Prof. Dr.-Ing. Gisela Lanza and Prof. Dr. Thomas S. Spengler for their valuable contributions to the finalization of this book.

Another great thank you goes to all my former colleagues. Over the years, many of you have become good friends. It was a pleasure to work with you in this exceptionally positive and motivating atmosphere. Especially, I want to emphasize my gratitude to Prof. Dr.-Ing. Sebastian Thiede as former head of the Sustainable Manufacturing research group. Your guidance, drive and ideas contributed significantly to the successful completion of this book. Further, special thanks go to Dr.-Ing. Denis Kurle for reviewing this book. Likewise, I want to acknowledge all external colleagues with whom I was allowed to work together.

Moreover, I want to thank my friends for their support and patience. Heartfelt thanks go to my parents Irene and Gerd Blume for their unwavering support and the opportunities they have opened up for me. Finally, and most of all, I want to thank Christine, whose contribution to this book cannot be valued enough. I love you.

Braunschweig, Germany
October 2020

Stefan Alexander Blume

Contents

Acronyms

ABS	Agent-based simulation
AHP	Analytical hierarchy process
ANP	Analytical network process
BAT	Best available technology
BI	Business intelligence
BM	Benchmarking
BREF document	Best available technology reference document
CML	Centrum voor Milieukunde Leiden
CNC	Computer numerical control
CO_2, CO_{2eq}	Carbon dioxide, carbon dioxide equivalents
DES	Discrete event simulation
DI	Data interpretation
DIN	Deutsches Institut für Normung (German Institute for Standardization)
DS	Dynamic systems
DSS	Decision support system
EFS	Energy flow simulation
ELCD	European Reference Life Cycle Database
ERP	Enterprise resource planning
ES	Expert system
EVSM	Energy value stream mapping
exVSM	Extended value stream mapping
GHG	Greenhouse gas
GUI	Graphical user interface
IC	Integrated computational
IED	Industrial emission directive
ILCD	International Reference Life Cycle Data System
ISO	International Organization for Standardization
KBS	Knowledge-based system
KMS	Knowledge management system

LCA	Life cycle assessment
LCC	Life cycle costing
LCI	Life cycle inventory
LCIA	Life cycle impact assessment
LPC	Load profile clustering
MADM	Multi-attribute decision making
MCDA	Multi-criteria decision analysis
MEFA	Material and energy flow analysis
MEMAN	Integral Material and Energy Flow Management in Manufacturing Metal Mechanic Sector
MFA	Material flow analysis
MFCA	Material flow cost accounting
MFS	Material flow simulation
MJ, MJ_{eq}	Megajoule, megajoule equivalent
ML	Matching level
MODM	Multi-objective decision making
MP	Market price
MSS	Management support system
OECD	Organization for Economic Co-operation and Development
OEM	Original equipment manufacturer
OR	Operations research
pcs.	Pieces
PDCA	Plan-do-check-act
PG	Performance gap
PhO	Physical optimum
PI	Performance indicator
PROMETHEE	Preference ranking organization method for enrichment of evaluations
QVSM	Quality value stream mapping
R	Requirement
ROI	Return on investment
RQ	Research question
S	Scenario
S&FT	Simulation and forecasting technologies
SD	System dynamics
SETAC	Society of Environmental Toxicology and Chemistry
SF	Single factory
SLCA	Social life cycle assessment
SME	Small- and medium-sized enterprises
SQL	Structured query language
TBS	Technical building services
TCO	Total cost of ownership
TOPSIS	Technique for order preference by similarity to ideal solution
UNEP	United Nations Environment Programme
VBA	Value benefit analysis

VC	Value chain
VDI	Verein Deutscher Ingenieure (Association of German Engineers)
VDMA	Verband Deutscher Maschinen- und Anlagenbau (Mechanical Engineering Industry Association)
VSM	Value stream mapping
WEEE Directive	Waste Electrical and Electric Equipment Directive

Symbols

a_{ij}	Element of adjacency matrix in row i and column j
$\vec{a}_m$	Attribute vector for measure m
a_{c_d}	Measure sub-attribute d from attribute category c
a_p	Chip thickness (turning, milling)
A	Affluence
A_m	Set of attributes induced by measure
A_s	Set of attributes induced by improvement situation
A_p	Product surface size
AM	Adjacency matrix
AV_{mach}	Availability of machine
b_0, b_r	(Unknown) constants characterizing system model behavior
BAT_j	Best available technology value for process j
BS_{proc}	Batch size of process
c_p	Specific heat capacity of material (hardening)
CD_f	Customer demand of a factory
CED	Cumulative energy demand
CF_r	Cost factor per output unit
CF_w	Cost factor per input unit of resource w
CF_y	Cost factor per product unit
$CR_{p,\ EOL}$	Collection rate for a product in end of life stage
CT_{proc}	Cycle time of process
d	Die closed depth (die bending)
E	Edge
EC	Energy costs
EI, EI_j	Energy intensity (of a process j)
G	Digraph
GCT	Gross cycle time
G_{proc}	Process group acc. to DIN 8580
GWP	Global warming potential
ΔH_f^{chrome}	Net enthalpy of chrome formation (hard chrome plating)

$\Delta H_m(T_1)$	Melting enthalpy at initial temperature (injection molding)
$\Delta H_m(T_2)$	Melting enthalpy at target temperature (injection molding)
HR_{mach}	Machine hour rate
I	Impact
i	System component (process, TBS element)
ID_f	Factory ID
ID_{mach}	Machine ID
ID_p	Product ID
ID_{proc}	Process ID
$IF_{mm,\ EOL}$	Impact factors per material mass contained in product
IF_r	Impact factor per output unit
IF_w	Impact factor per input unit of resource w
j	System component (process, TBS element)
J	Variance
j_{crit}	Critical process
k	Number of clusters used for k-means clustering
$k_{c1.1}$	Specific cutting force (turning, milling)
K_L	Correction factor for workpiece shape (turning, milling)
K_{TM}	Correction factor for tool material (turning, milling)
K_{TW}	Correction factor for tool wear (turning, milling)
K_V	Correction factor for cutting speed (turning, milling)
K_{form}	Constant for die opening distance (die bending)
K_γ	Correction factor for cutting angle (turning, milling)
l	Length of bend (die bending)
LC	Labor costs
LIT_p	Life time of a product
LOC_f	Factory location
LT	Lead time
μ_i	Cluster average values
m	Mass of workpiece (hardening)
m_i	Improvement measure
m_c	Chip thickness exponent (turning, milling)
m_{chrome}	Mass of deposited chrome (hard chrome plating)
m_p	Product mass
M	Set of alternative improvement measures
M_{chrome}	Molar mass of chrome (hard chrome plating)
MC	Material costs
ML	Matching level between measure and improvement situation
$mm_{p,\ EOL}$	Mass of material contained in a product
MUE	Material utilization efficiency
MUR	Machine utilization rate
NCT	Net cycle time
N_f	Factory name
N_{mach}	Machine name

NF_{m_i}	Net flow for measure m_i
NOM_{proc}	Number of machines working in parallel
NOW_{mach}	Number of workers per machine
N_p	Product name
N_{proc}	Process name
$NVAEI$	Non-value adding energy intensity
OC	Other costs
OEE	Overall equipment effectiveness
$OF^-_{m_i}$	Negative outranking flows for measure m_i
$OF^+_{m_i}$	Positive outranking flows for measure m_i
OG, OG_j	Organizational performance gap
$OP_{p,\ EOL}$	End of life option for a product
$\vec{p}, \vec{p}_m, \vec{p}_{init}, \vec{p}_{change}$	Performance vectors
P	Population
PD_{mach}	Product demand of machine
PhO_j	Physical optimum value for process j
PS	Performance score
PT_{proc}	Processing time
σ_m	Ultimate tensile strength (die bending)
$QDR_{p,\ EOL}$	Quality degradation rate
QR_{proc}	Quality rate of process
r_j	Waste flows from j leaving the system
RIF_j	Resource intensity factor of TBS element j
RO_{mach}	Resource demand level of machine in off state
RP_{mach}	Resource demand level of machine in processing state
$RR_{p,\ EOL}$	Recycling rate in end of life stage
RR_{proc}	Rework rate of process
RS_{mach}	Resource demand level of machine in standby state
RW_{mach}	Resource demand level of machine in waiting state
SCS_{proc}	Successor of process
SEC, SEC_j	Specific energy consumption
$\vec{s}$	Improvement situation vector
s_{c_d}	Improvement situation attribute
s_j	Stock changes of j due to imbalances in entering and leaving flows
S_i	Cluster
t^2_{wp}	Stock thickness (die bending)
T	Technology factor
T_1	Initial temperature (hardening)
T_2	Target temperature (hardening)
TC	Total costs
TCG, TCG_j	Technical performance gap
TLG, TLG_j	Technological performance gap
TPG, TPG_j	Total performance gap

T_{proc}	Process type acc. to DIN 8580
TT	Customer takt time
V	Vertice
V_{chip}	Removed chip volume (turning, milling)
$VAEI$	Value adding energy intensity
$VASEC, VASEC_j$	Value adding specific energy consumption
$\vec{w}$	Weighting vector
w_i	Individual weight of PI_i
w_j	Resource inputs from outside the system boundaries to j
$w_{j,\,energy}$	Energy inputs from outside the system boundaries to j
$w_{j,\,material}$	Material inputs from outside the system boundaries to j
w_{kj}	Resource input flows from a source k outside the system boundaries to j
W_j	Aggregated resource demand of TBS element j
WD_f	Working days per year in a factory
WH_f	Working hours per day in a factory
x_j	Metered values used for k-means clustering
X_{crit}	Critical parameter
X_r	Model input parameters
y_j	Usable product flows from j leaving the system
Y_i	Model outputs
$z_{ij},\, z_{ji}$	Intermediate flows between j and another system element i
z_{jj}	Intermediate flows looping from j to j
Z_j	Aggregated provided intermediate flow quantity of TBS element j

Chapter 1
Introduction

In this introduction, the general need to increase resource efficiency in manufacturing value chains is outlined, substantiating the demand for adequate decision support in Sect. 1.1. Based on this, research questions are derived and the structure of this work is introduced in Sect. 1.2.

1.1 Motivation and Problem Statement

Manufacturing deals with the creation of physical products to satisfy human needs using various manufacturing processes, which alter shape and physical properties of input materials (Chryssolouris 2006; Westkämper 2006). Manufacturing is a main driver for economic development and has a large positive impact on the welfare of nations (Gutowski et al. 2013). However, the extent of present manufacturing activities results in enormous global energy and material flows, causing harm to both local and global environment as well as humans as part of this ecosphere. According to the definition of a sustainable development (compare Fig. 1.1), industrial activities should thus not only focus on economic targets but also consider social and environmental consequences (World Commission on Environment and Development 1987; Elkington 1998).

Historically, the focus of industrial activities has often been put on the economic dimension, while environmental impacts have hardly been considered (Pearce and Atkinson 1993; von Hauff and Wilderer 2008; Rockström 2015). As a consequence, today's global consumption rates for many resources and related harmful emissions are beyond the natural regeneration rate of our planet and current industrial activities must be regarded as unsustainable. This is underlined by the *Earth Overshoot Day*, which marks the day that humanity's environmental footprint exhausts the

© The Author(s), under exclusive license
to Springer Nature Switzerland AG 2020
S. A. Blume, *Resource Efficiency in Manufacturing Value Chains*,
Sustainable Production, Life Cycle Engineering and Management,
https://doi.org/10.1007/978-3-030-51894-3_1

Fig. 1.1 Target triangle of sustainable development with influencing factors. Adapted from von Hauff and Wilderer (2008) and Herrmann (2010)

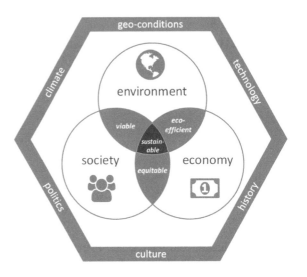

Earth's annual regenerative capacity (Posthuma et al. 2014). Accordingly, the growth of global resource utilization and related environmental damage progresses with alarming speed, resulting in a total human environmental footprint of 1.7 earth's in 2019 (Global Footprint Network 2018). As economy and society can only grow and sustain by respecting the natural carrying capacity of the earth, existing planetary boundaries should be better respected. Accordingly, there is a strong need to shift future priorities towards the environmental dimension of sustainability (Rockström et al. 2009; Hauschild 2015; Rockström 2015). Although several measures to reduce the environmental impacts of industry have been implemented, positive effects have so far been overcompensated by other factors on a global level. By means of the *IPAT equation* (compare Eq. (1.1)), a basic explanation can be provided. It describes the human impact on the environment *I* as a mathematical product of the size of human population *P*, human affluence *A*, i.e. the value created or consumed per capita, and a technology factor *T*, which describes the environmental impacts per created value (Commoner 1971; Ehrlich and Holdren 1971; Chertow 2000).

$$I = P * A * T \qquad (1.1)$$

While great improvements have been achieved with regard to the technology factor, population and affluence have significantly increased over the last decades and are estimated to grow further. Altogether, technological improvements have obviously been overcompensated by population growth and increasing affluence including rebound effects (Berkhout et al. 2000; Figge et al. 2014). These findings necessitate to rethink the general underlying principles of industrial value creation, promoting the use of renewable resources and closing of energy and material loops (Hauschild et al. 2017; Kara et al. 2018).

Today, for many industrial actors an economic pressure still constitutes the main incentive to use resources efficiently. This pressure will probably intensify due to an ongoing increase in resource costs and cost volatilities (Dobbs et al. 2011; de Groot et al. 2012). In recent years, growing customer demands for environmentally sustainable products and legislative pressure have also started to motivate companies to act more environmentally sustainable (European Union 2000, 2012; Manget et al. 2009). As a consequence, companies are more and more trying to improve their eco-efficiency by managing environmental and economic performances simultaneously. Despite all efforts and research carried out in this field, several barriers still prevent industry to exploit its full resource saving potential. These comprise a limited awareness about improvement potentials, a limited availability of resources to conduct thorough analyses, a lack of available expert knowledge, long payback periods, missing knowledge about environmental consequences of actions or trade-offs between different target dimensions and performance indicators (Foran et al. 2005; Thollander 2008; Bey et al. 2013; Mittal et al. 2013). Due to complex and dynamic cause-effect relations between elements in manufacturing systems like human actors, production machines or technical building services, companies often struggle with the translation of general improvement strategies into specific improvement measures (Sproedt et al. 2015). A main barrier can also be seen in the international organization of today's economy. Through the establishment of global value chains and supply networks during the last decades, improvement initiatives focusing local entities of manufacturing systems like single processes in an isolated manner are hardly sufficient to achieve significant global improvements (Brondi et al. 2014). Global consequences can result from diverse local decisions, e.g. changes in product design, resource supply, process and technology selection or location planning. These consequences can be either positive or negative and they may lead to reduced or increased costs or environmental impacts (compare Fig. 1.2). Often, consequences are ambiguous, comprising both positive and negative effects on different scales from local to global, e.g. if an economic improvement is achieved at the expense of greater

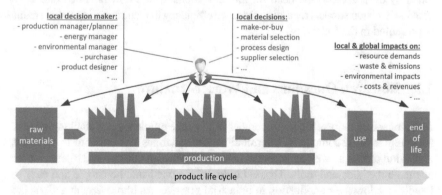

Fig. 1.2 Potential effects of local decision making on other value chain actors and in other product life cycle stages

environmental impacts or vice versa. Unfortunately, local improvements in industrial countries do also contribute to reinforce global inequalities: while consumers and producers in high-income countries profit from global economic growth, low-income countries often bear negative environmental or social consequences (Hubacek et al. 2016). Therefore, it is more than ever necessary to identify and implement practices that provide competitive advantages on value chain level but also entail improvements on local level (Green et al. 2008). For the sake of overall improvements, local negative consequences can in some cases be tolerable, but may necessitate compensative mechanisms. As indicated in Fig. 1.2, one major challenge arises from the need to expand the scope of activities to all phases of product life cycles (from cradle to grave) instead of focusing on the production stage only (Herrmann 2010; Walther 2010).

The growing complexity of today's value chain structures necessitates cross-actor cooperation to analyze and improve both local and global economic and environmental consequences of industrial activities. A holistic perspective on the entire value chain and an active product life cycle management are prerequisites to achieve significant improvements while avoiding problem shifting in a spatial or timely sense (Herrmann 2010). This also goes along with growing challenges in terms of system design and operation (Walther 2010). Adequate methods and tools are needed to assess alternative courses of action and to balance resulting consequences in case of target conflicts not only within a single company but along a value chain or product life cycle. Decision makers should be empowered to better understand system inherent cause-effect relations, e.g. to quantify the effects of using other resources, applying different process parameters or using other production technologies. Due to the large number of system elements and possible influencing factors, methods and tools for system modeling and improvement must follow a holistic perspective on manufacturing. This means to take into account different spatial and temporal scopes, e.g. from single processes up to value chain level or product life cycle and from operational up to strategic planning (Herrmann et al. 2011; Thiede 2012). Motivated by these factors, the need for and development of a new holistic concept to provide decision support regarding resource efficiency in manufacturing value chains is presented in the following.

1.2   Research Questions and Work Structure

As emphasized before, the improvement of resource efficiency is a challenging task for manufacturing companies. Interrelations and dynamics in manufacturing systems and value chains as well as potential trade-offs between competing targets call for adequate methods and tools. These should, on the one hand, be simple and easy to handle to allow their application in industrial practice. As improvement actions are typically expected to pay off in an economic sense, efforts to run an analysis and related costs are highly critical. If methods and tools are too costly, complex or time consuming to apply, they might be hardly useful in practice. On the other hand, they must adequately preserve the complexity of real-world systems to ensure the validity of generated results. Taking this into consideration together with the developments and challenges described in Sect. 1.1, two main research questions can be formulated (compare Table 1.1).

In order to answer these research questions, the state of the art regarding resource efficiency in manufacturing value chains is first described in Chap. 2. Based on a general introduction into the topic of industrial production, the need for a resource efficient manufacturing is discussed. A focus is put on well-established methods to model manufacturing systems in terms of resource efficiency. Furthermore, an introduction into decision theory is provided in order to better understand the requirements of decision makers in the context of manufacturing. A necessity to turn special attention to the applicability of developed methods and tools is emphasized. Chap. 3 is dedicated to a review of the state of research regarding resource efficient manufacturing. Criteria are derived from the research questions and applied to existing work in order to judge about their suitability to tackle the challenges presented ahead. A research demand is identified by a comparison of existing approaches. It serves as basis to develop a new concept to increase resource efficiency in manufacturing value chains as presented in Chap. 4. Therefore, specific requirements are defined, addressing the gaps identified in existing work. The new concept is then described in detail from a rather theoretical perspective following a stepwise improvement procedure. In Chap. 5, the practical implementation of the concept into a software demonstrator is outlined with a focus on the applicability from user perspective. Potentials and flexibility of the developed concept are demonstrated by means of two industrial case studies in Chap. 6, which highly differ in the underlying requirements. Finally, Chap. 7 summarizes tackled challenges as well as developed solutions and gives an outlook on potential future research. The work structure is summarized in Fig. 1.3.

Table 1.1 Research questions

RQ 1	How to *holistically assess resource efficiency* in manufacturing value chains?
	This question refers to both already existing methods and tools but also new combinations, extensions or entirely new options to model and assess manufacturing systems. They should therefore preserve a certain degree of system complexity and allow to consider technical, economic and environmental issues. Broad spatial and temporal scopes should be covered in order to avoid problem shifting, e.g. by considering multiple system levels of manufacturing as well as product life cycle stages.
RQ 2	How to provide *decision support* for system improvement, respecting limited resources such as time and expert knowledge?
	This question addresses the need to conduct thorough analyses of complex systems and to find suitable improvement options with a limited availability of resources. Therefore, possibilities to provide decision support and to reduce efforts for time consuming and knowledge intensive activities need to be identified and developed. Such support may for instance aim at the identification of improvement measures or the handling of potential target conflicts resulting from alternative courses of action.

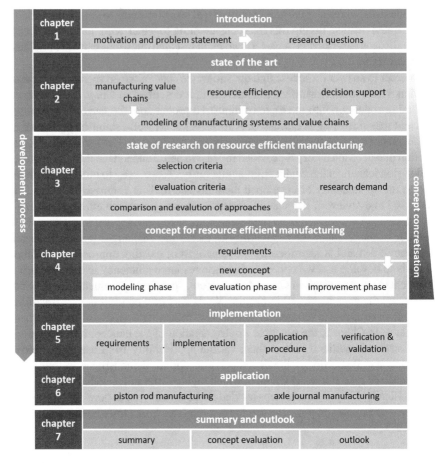

Fig. 1.3 Overall structure of book

References

Berkhout PHG, Muskens JC, Velthuijsen JW (2000) Defining the rebound effect. Energy Policy 28(6–7):425–432

Bey N, Hauschild MZ, McAloone TC (2013) Drivers and barriers for implementation of environmental strategies in manufacturing companies. CIRP Ann 62(1):43–46. https://doi.org/10.1016/j.cirp.2013.03.001

Brondi C, Fragassi F, Pasetti T, Fornasiero R (2014) Evaluating sustainability trade-offs along supply chain. In: International conference on engineering, technology and innovation: engineering responsible innovation in products and services

Chertow MR (2000) The IPAT equation and its variants. J Ind Ecol 4(4):13–29. https://doi.org/10.1162/10881980052541927

Chryssolouris G (2006) Manufacturing systems: theory and practice, 2nd edn. Springer, New York

Commoner B (1971) The closing circle: nature, man, and technology. Bantam Books Inc., New York de Groot HLF, Rademaekers K, Svatikova K et al (2012) Mapping resource prices: the past and the future. ECORYS Nederland BV, Rotterdam

Dobbs R, Oppenheim J, Thompson F, et al (2011) Resource revolution: meeting the world's energy, materials, food, and water needs. McKinsey Global Institute

Ehrlich PR, Holdren JP (1971) Impact of population growth. Science (80-) 171(3977):1212–1217. https://doi.org/10.1126/science.171.3977.1212

Elkington J (1998) Partnerships from cannibals with forks: the triple bottom line of 21st-century business. Environ Qualty Manag 8(1):37–51

European Union (2012) Directive 2012/19/EU of the European Parliament and of the Council of 4 July 2012 on waste electrical and electronic equipment (WEEE)

European Union (2000) Directive 2004/35/CE of the European Parliament and of the Council of 21 April 2004 on environmental liability with regard to the prevention and remedying of environmental damage

Figge F, Young W, Barkemeyer R (2014) Sufficiency or efficiency to achieve lower resource consumption and emissions? The role of the rebound effect. J Clean Prod 69:216–224. https://doi.org/10.1016/j.jclepro.2014.01.031

Foran B, Lenzen M, Dey C, Bilek M (2005) Integrating sustainable chain management with triple bottom line accounting. Ecol Econ 52(2):143–157. https://doi.org/10.1016/j.ecolecon.2004.06.024

Global Footprint Network (2018) Past earth overshoot days. https://www.overshootday.org/newsroom/past-earth-overshoot-days/. Accessed on 10 Aug 2019

Green Jr KW, Whitten D, Inman RA (2008) The impact of logistics performance on organizational performance in a supply chain context. Supply Chain Manag 13(4):317–327. https://doi.org/10.1108/13598540810882206

Gutowski TG, Allwood JM, Herrmann C, Sahni S (2013) A global assessment of manufacturing: economic development, energy use, carbon emissions, and the potential for energy efficiency and materials recycling. Annu Rev Environ Resour 38(1):81–106. https://doi.org/10.1146/annurev-environ-041112-110510

Hauschild MZ (2015) Better – But is it good enough? on the need to consider both eco-efficiency and eco-effectiveness to gauge industrial sustainability. Procedia CIRP 29:1–7. https://doi.org/10.1016/j.procir.2015.02.126

Hauschild MZ, Herrmann C, Kara S (2017) An integrated framework for life cycle engineering. Procedia CIRP 61:2–9. https://doi.org/10.1016/j.procir.2016.11.257

Herrmann C (2010) Ganzheitliches life cycle management. Springer, Heidelberg, Dordrecht, London, New York

Herrmann C, Thiede S, Kara S, Hesselbach J (2011) Energy oriented simulation of manufacturing systems – Concept and application. CIRP Ann 60(1):45–48. https://doi.org/10.1016/j.cirp.2011.03.127

Hubacek K, Feng K, Chen B, Kagawa S (2016) Linking local consumption to global impacts. J Ind Ecol 20(3):382–386. https://doi.org/10.1111/jiec.12463

Kara S, Hauschild MZ, Herrmann C (2018) Target-driven life cycle engineering: staying within the planetary boundaries. Procedia CIRP 69:3–10. https://doi.org/10.1016/j.procir.2017.11.142

Manget J, Roch C, Munnich F (2009) Report capturing the green advantage for consumer companies. The Boston Consulting Group, Boston

Mittal VK, Egede P, Herrmann C, Sangwan KS (2013) Comparison of drivers and barriers to green manufacturing: a case of India and Germany. Re-Eng Manuf Sustain 723–728. https://doi.org/10.1007/978-981-4451-48-2_118

Pearce D, Atkinson G (1993) Capital theory and the measurement of sustainable development: an indicator of "weak" sustainability. Ecol Econ 8:103–108

Posthuma L, Bjørn A, Zijp MC et al (2014) Beyond safe operating space: finding chemical footprinting feasible. Environ Sci Technol 48(11):6057–6059. https://doi.org/10.1021/es501961k

Rockström J (2015) Bounding the planetary future: why we need a great transition. Gt Transit Initiat 9:1–13

Rockström J, Steffen WL, Noone K, Al E (2009) Planetary boundaries: exploring the safe operating space for humanity. Ecol Soc 14(2):81–87. https://doi.org/10.1007/s13398-014-0173-7.2

Sproedt A, Plehn J, Schönsleben P, Herrmann C (2015) A simulation-based decision support for eco-efficiency improvements in production systems. J Clean Prod 105:389–405. https://doi.org/10.1016/j.jclepro.2014.12.082

Thiede S (2012) Energy efficiency in manufacturing systems. Springer, Berlin, Heidelberg

Thollander P (2008) Towards increased energy efficiency in swedish industry – Barriers, driving forces & policies. Linköping University

von Hauff M, Wilderer PA (2008) Industrial ecology: engineered representation of sustainability. Sustain Sci 3(1):103–115. https://doi.org/10.1007/s11625-007-0037-6

Walther G (2010) Nachhaltige Wertschöpfungsnetzwerke. Gabler, Wiesbaden

Westkämper E (2006) Einführung in die Organisation der Produktion. Springer, Berlin, Heidelberg

World Commission on Environment and Development (1987) Report of the World Commission on Environment and Development: Our Common Future

Chapter 2
Resource Efficiency in Manufacturing Value Chains

This chapter presents general background information about resource efficiency in manufacturing value chains. In Sect. 2.1, a general introduction into industrial production and manufacturing with its inherent structure, challenges and meaningful indicators for performance measurement is provided. The term of resource efficiency in the context of manufacturing is then described in Sect. 2.2, emphasizing suitable improvement strategies. In Sect. 2.3, the relevance of decision support for a more resource efficient manufacturing is substantiated and suitable methods and tools for decision making are discussed. Finally, an overview about well-established methods to model manufacturing systems is given in Sect. 2.4.

2.1 Manufacturing Value Chains

2.1.1 Production as Transformation Process

Industrial production bases on value creation, i.e. on generating value-added outputs (products and services) by using material or immaterial resource inputs (Günther and Tempelmeier 2005). From an economic perspective, highly relevant input resources are typically human labor, material and energy (European Commission – Joint Research Centre 2016), including raw materials, consumables, semi-finished products as well as different energy forms like electricity, gas or fuel (Rebhan 2002). An exemplary transformation process from input resources into desired and undesired products such as waste and emissions is indicated in Fig. 2.1. In a broader sense, the term *production* is not limited to the direct creation of products, but can also be

© The Author(s), under exclusive license
to Springer Nature Switzerland AG 2020
S. A. Blume, *Resource Efficiency in Manufacturing Value Chains*,
Sustainable Production, Life Cycle Engineering and Management,
https://doi.org/10.1007/978-3-030-51894-3_2

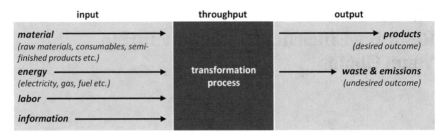

Fig. 2.1 Production as transformation process, adapted from Dyckhoff and Spengler (2010) and Schenk et al. (2014)

understood as the sum of technical and organizational processes to create, sustain and recycle products over their whole life cycles (Westkämper 2006).

2.1.2   Structure of Manufacturing Systems

Manufacturing as "transformation of materials and information into goods for the satisfaction of human needs" can be regarded as a sub-process of production (Chryssolouris 2006). It refers to the creation of physical products using manufacturing processes, which alter shape and physical properties of given input materials (Westkämper 2006). In order to conduct these transformation processes, manufacturing equipment is needed. The superior socio-technical system of processes, equipment and humans is referred to as *manufacturing system* (Chryssolouris 2006). In order to better describe and analyze manufacturing systems, different vertical hierarchy levels can be distinguished depending on their inherent elements, functions, interfaces, inputs and outputs. A popular classification comprises – from lowest to highest level – processes, workstations, production groups, production areas, factories, production sites and production networks (Wiendahl et al. 2005; Westkämper 2006). Various other authors propose similar classifications (Herrmann 2010; Herrmann et al. 2010, 2014; Reich-Weiser et al. 2010; Verl et al. 2011; Duflou et al. 2012; Heinemann 2016; Schönemann 2017). The distinction of at least four levels seems to be meaningful according to most taxonomies (compare Fig. 2.2).

> **Process**: This level comprises the conduction of manual and machine-based manufacturing processes by converting resources into products (compare Fig. 2.1). The relation between machines and processes is quite close, as a processes capabilities and limitations highly depend on the design and operation of the machine that performs it (Chryssolouris 2006). For discrete manufacturing, i.e. the production of distinct products, typical manufacturing processes can be found in DIN 8580 (Deutsches Institut für Normung 2003).

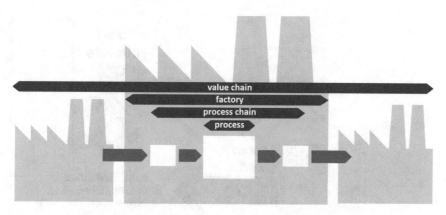

Fig. 2.2 Hierarchical structure of manufacturing systems

Process chain: This level features several interlinked manufacturing processes. Manufacturing process chains can be structured in different ways regarding material flow and spatial layout of manufacturing equipment, e.g. as job shop, cellular production or line production (Schenk et al. 2014). Accordingly, the link between processes may be either permanently established or lose, i.e. individual for a product type or even every single workpiece. Further distinctions can be made between continuous, diverging, converging and circular flows (Dyckhoff and Spengler 2010; Heinemann 2016), while the latter features the highest degree of complexity (compare Fig. 2.3).

Factory: This level refers to a combination of process chains with other constituent factory elements (Thiede 2012). In order to manage the operation of such an entire factory system, a holistic factory model is helpful. As illustrated in Fig. 2.4, complex and dynamic interrelations between the main factory elements – production machines, technical building services (TBS), building shell – and

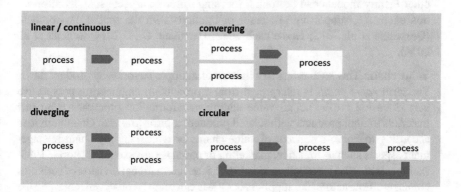

Fig. 2.3 Types of product flows in manufacturing systems, adapted from Dyckhoff and Spengler (2010) and Heinemann (2016)

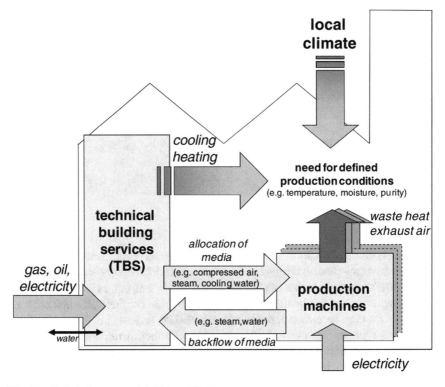

Fig. 2.4 Holistic factory model (Thiede 2012)

their physical exchanges of material, energy and media need to be considered. Further, factories are embedded into their local surroundings. Thus, local conditions such as climate conditions often play an important role for factory operation and must be considered. Apart from this factory representation, multiple other factory models can be found, focusing different aspects such as information exchange, supporting processes or relations with the spatial environment (Despeisse et al. 2012; Hesselbach 2012; Herrmann et al. 2014; Schenk et al. 2014).

Value chain: This level refers to a linkage of multiple factories by product flows. The term *value chain* is interpreted from a production engineering perspective here, focusing on the actual value adding transformation processes from raw materials to final products with related production infrastructure. Other activities according to Porter's concept of value chains or typical supply chain management aspects such as logistics, marketing or purchase are excluded (Porter 1985; Mentzer et al. 2001). As indicated in Fig. 2.5, an international division of activities among multiple actors like suppliers or OEMs can often be observed, entailing the establishment of global value chains or networks (Günther and Tempelmeier 2005; Westkämper 2006). The challenge to manage and improve such cross-company

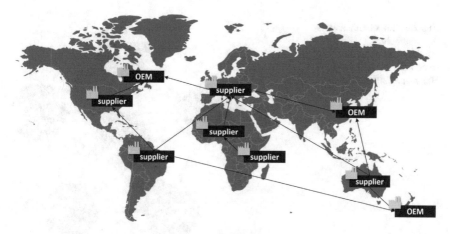

Fig. 2.5 Example for network of industrial actors on value chain level

networks is going far beyond the management of a single factory. On the one hand, multiple criteria such as costs, delivery times, quality, flexibility, risks or market access need to be taken into account to find a satisfying value chain configuration (Lanza and Ude 2010). On the other hand, an effective and efficient value chain operation requires a strong cooperation of industrial actors as well as adequate supporting tools (Walther 2010).

2.1.3 Target Dimensions and Performance Measurement

Managing manufacturing companies means to deal with multi-dimensional target systems. These dimensions comprise – from traditional to rather novel – costs, time, quality, flexibility and environment (Günther and Tempelmeier 2005; Chryssolouris 2006; Herrmann 2010), while often a balancing between more than one dimension is needed. Therefore, a quantification of target achievements is mandatory to derive meaningful decisions. Challenges arise due to inherent interrelations and occurring trade-offs between the target dimensions as indicated in Fig. 2.6, whereby a simultaneous optimization of all dimensions is often not possible (Chryssolouris 2006; Größler and Grübner 2006). However, also synergetic improvements addressing multiple target dimensions are achievable. As an example, improving quality, reliability and delivery time is regarded as a catalyst for long-term cost savings (Ferdows and De Meyer 1990).

In order to manage manufacturing systems, the conduction of performance measurement is inevitable. It allows managers to drive a target-oriented development of the organization and adapt to changing external requirements by comparing targets and measured performance (Beer 1995). The underlying theory of controlling using feedback loops is known as cybernetics (Wiener 1948). This approach necessitates the definition of performance indicators (PIs). They aim to indicate

Fig. 2.6 Target dimensions of manufacturing with their relations and potential trade-offs, adapted from Herrmann (2010)

quantitative measurable relations in a condensed way, i.e. in a simple but still precise form (Dyckhoff and Spengler 2010). Thus, they should fulfil certain criteria such as objectivity, relevance, comprehensibility or validity (Brown 1997). To give an example, the return on investment (ROI) is one important economic PI to express the financial profitability of business activities. The indicator bases upon a greater performance indicator network, in which the single PIs are linked by mathematical operations (Staehle 1969; Botta 1997).

Economic targets may be the core of most business activities. However, economic PIs are directly or indirectly influenced by other events and activities (Macharzina and Wolf 2008; Westkämper and Zahn 2008). Consequently, a complete formulation of targets based on economic PIs is not regarded as sufficient to manage a company. Traditional PIs of production concerning quality and time aspects like machine availability rate, quality rate, lead time or overall equipment effectiveness (OEE) should also be used for decision making (Muchiri and Pintelon 2008; Schenk et al. 2014). The need to incorporate environmental PIs is increasingly emphasized in recent years (Günther 1994; Haag 2013; Fantini et al. 2015) and is also subject of international standards like ISO 14031 (International Organization for Standardization 2013). Resource flow related PIs are of particular interest, as they are linked to both costs but also environmental impacts. A wide range of corresponding PIs is for instance shown by Fantini et al. (2015) and Heinemann (2016). Regarding energy efficiency, typical PIs can be found in Patterson (1996), Bunse et al. (2011), Haag (2013), Zein (2013) and Posselt (2016). A popular example to measure the energy related performance on process level is the calculation of a specific energy consumption (SEC), relating the energy demand for a specific process to a reference such as a processed mass or volume (compare Table 2.1). An aggregation of SECs on process chain or factory level can be used to express the overall energy intensity (EI) to manufacture a certain product (Dehning et al. 2017).

Table 2.1 Typical references for SEC calculations depending on process main groups according to DIN 8580, adapted from Erlach and Westkämper (2009)

Reference	Main groups	Processes (examples)
Mass of processed material	1 Primary shaping 2 Reshaping 6 Changing material properties	Die casting, injection molding Hot rolling, deep drawing Thermal treatment
Mass of removed material	3 Separating	Grinding, milling, turning
Contact area/length	4 Joining	Welding, bonding
Processed surface size	5 Coating	Painting, electroplating

2.2   Resource Efficiency

2.2.1   Sustainable Manufacturing

Most concepts and terms which link sustainability with manufacturing have in common that they promote life cycle management, i.e. the management of entire product life cycles (Hauschild et al. 2017). As indicated in Fig. 2.7, this typically refers to the four main life cycle stages raw material extraction, production, use and end of life (e.g. recycling or disposal) as well as potential intermediary activities such as transports (Klöpffer and Grahl 2009). This understanding is also an underlying concept of established methods like life cycle costing (compare Sect. 2.4.4) and life cycle assessment (compare Sect. 2.4.5). A sustainable manufacturing addresses in particular the production stage of the product life cycle without neglecting the economic and environmental consequences of activities in other life cycle stages. It can also be defined as "efficient conversion of resources into value for society" (Despeisse 2013), emphasizing the need to address all three dimensions of sustainability. Further, it should take a holistic perspective on manufacturing by

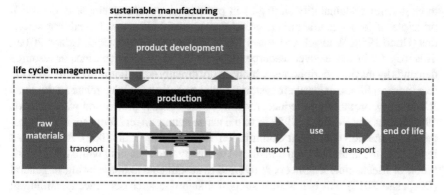

Fig. 2.7 Typical life cycle of a tangible product, adapted from Herrmann (2010), Ohlendorf (2006) and Klöpffer and Grahl (2009)

considering different system levels and applying different sustainability strategies (Herrmann et al. 2008). A sustainable manufacturing should also aim to decouple economic growth from related negative environmental impacts (Braungart et al. 2007; Herrmann et al. 2007; Hauschild 2015). Modern approaches typically promote a *strong sustainability* approach, i.e. they grant highest priority to the conservation of the Earth's life support system before considering the interests of society and economy (Hauschild et al. 2017).

Sustainability in manufacturing can be achieved in different ways. Typically, three general strategies are distinguished:

- An *efficiency strategy* can be described as "doing more with less", i.e. either minimizing resource inputs, maximizing product outputs or combining both principles (Huber 1994; Herrmann 2010). The strategy promotes to improve existing manufacturing systems in a rather evolutionary way, e.g. by running machines in their ideal working points (Schudeleit et al. 2016). A focus on efficiency strategies is increasingly criticized as "damage management" and "dead end" (Despeisse et al. 2013; Herrmann et al. 2014), as it does not allow for a real decoupling of economic growth and environmental impacts.
- An *effectiveness strategy* in contrast promotes to close the loop of technical and natural metabolism (Dyckhoff and Souren 2007). By substituting the use of harmful and non-renewable resources, this approach is rather revolutionary. Research is promoting to shift the focus towards effectiveness strategies in order to become sustainable in absolute but not only relative terms (Braungart et al. 2007; Hauschild 2015).
- A *sufficiency strategy* promotes to consume less by changing human consumption patterns. It is therefore rather an instrument for consumers than for producers. However, sufficiency strategies can also be applied to reduce resource demands in manufacturing systems, e.g. by shutting down components or entire machines in non-production periods (Schudeleit et al. 2016).

More concrete principles for manufacturing have been derived based on these three general sustainability strategies in practice. A popular example is the *muda* principle of lean manufacturing, promoting to avoid all non-value adding activities (Ohno 1988; Womack and Jones 1996; Rother and Shook 1999; Erlach 2010). This may for instance cover resource demands in non-productive times or resource demands by machine peripheries, which are not directly contributing to value creation (e.g. pumps, filters, exhaust air system). Other popular principles relate to the *three R's strategy*, promoting to reduce, remanufacture or reuse production waste (Sarkis 1995; Sarkis and Rasheed 1995), or to the *waste management hierarchy* of the European Commission (European Commission 1975; Hansen et al. 2002). As indicated in Fig. 2.8, the latter has been transferred to energy management (Wolfe 2005; Institution of Mechanical Engineers 2009) and to sustainable manufacturing in general (Despeisse 2013). All hierarchies suggest a stepwise approach, ideally aiming at avoidance as optimal solution. Overviews about proven and more specific approaches to improve sustainability on different manufacturing system levels can be found in

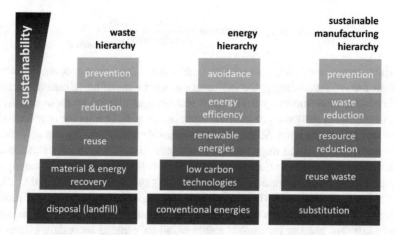

Fig. 2.8 Improvement hierarchies for waste, energy and sustainable manufacturing. Adapted from Despeisse (2013)

Duflou et al. (2012) and Despeisse (2013). Yet, a main challenge in practice can still be seen in the identification of applicable improvement measures for a specific case.

2.2.2 Resource Efficient Manufacturing

In both economic and environmental terms, an efficient use of resources is advisable. Resource efficiency as indicator to measure the use of resource demands in transformation processes is often used synonymously with related terms like resource productivity or resource intensity, while their exact definition depends on context and purpose (Diekmann et al. 1999; Organisation for Economic Cooperation and Development 2001). In this work, resource productivity is understood as the ratio of (wished) outputs and required inputs of a transformation process (Dyckhoff and Spengler 2010), for instance a product's mass after machining divided by the raw material mass needed as input:

$$resource\ productivity = \frac{wished\ outputs}{required\ inputs} \tag{2.1}$$

As indicated in Fig. 2.9, the resource productivity as the relation between input and output of a transformation process can be described as production function. By comparing different production functions, the degree of efficiency of a transformation can be assessed. In this context, resource efficiency can be understood as the ratio of this productivity and the productivity of a process reflecting best practice (Organisation for Economic Cooperation and Development 2001):

$$resource\ efficiency = \frac{productivity_{actual}}{productivity_{bestpractice}} \qquad (2.2)$$

Thus, resource efficiency is a relative indicator, taking into account the current state of the art. Accordingly, the indicator can give a quantitative estimation about achievable improvements by using latest technologies. Based on these definitions, improvements of resource productivity and efficiency indicators for transformation processes go hand in hand. Strategies for sustainable manufacturing can give indications about how to reach improvements (compare Sect. 2.2.1). However, following such strategies does often affect technical indicators like availability or throughput of manufacturing processes and entire manufacturing systems (Helu et al. 2011). These effects can be both synergetic but also contrary regarding other targets of production, i.e. time, quality, costs and flexibility (compare Sect. 2.1.3). As an example, energy efficient process parameters might entail unintended secondary effects like higher maintenance costs (Chryssolouris 2006; Schlosser et al. 2011). In this context, several researchers discussed the particular interrelations of *lean* and *green* manufacturing strategies (Herrmann et al. 2008; Diaz-Elsayed et al. 2013; Dües et al. 2013; Weinert et al. 2013; Greinacher et al. 2015; Fischer 2017; Abreu et al. 2017). Dües et al. (2013) concluded that these two paradigms are not incompatible but can be synergistically connected in many cases. A holistic (i.e. multi-criteria) analysis of measures as described in Sect. 2.3.3 is recommended to judge about the overall benefits and drawbacks in terms of sustainability (Weinert et al. 2013).

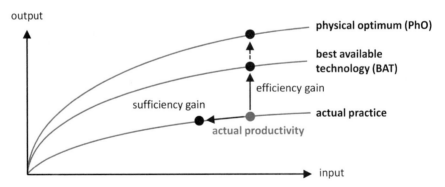

Fig. 2.9 Comparison of actual production function with best practice and optimal production function, adapted from Cantner et al. (2007) and Schmidt (2007)

2.3 Decision Support

2.3.1 Relevance of Decision Making for Manufacturing

The ability to make clear-cut decisions is regarded as number one management skill in business (Turban et al. 2007). Due to a growing complexity of business and its environment, decision making becomes more and more challenging. Turban et al. (2007) name some of the main drivers for this observation:

- a rapid development of (information) technology provides more alternatives to choose from
- increases in structural complexity and competition entail higher error costs and thus exacerbate the consequences of bad decisions
- higher uncertainties can be observed, e.g. through advancing globalization

All these developments also increase the need to make decisions even quicker. It can be stated, that more than ever it is impossible to rely on trial-and-error management approaches. Instead, there is a need for systematic approaches as well as effective tools to support decision making in industrial practice.

A *manufacturing decision* can be regarded as a selection of values for decision variables, i.e. a decision for one or another alternative course of action (Turban et al. 2007). This is typically related to the design or operation of manufacturing processes, machines or systems (Chryssolouris 2006). Decisions can relate to different planning horizons. According to Dyckhoff (2006) and Dyckhoff and Spengler (2010), three time horizons can be distinguished in production management: operational, tactical and strategic (compare Fig. 2.10). The operational time horizon typically addresses decisions made within a year, ranging from production scheduling over material disposition up to production sequence planning. A tactical time horizon usually stretches from one to five years, including tasks like the definition of manufacturing depth, capacity and layout planning or technology management. Finally, the strategic horizon refers to decisions with a forerun of more than five years, covering general target definitions, research and development activities or production location planning.

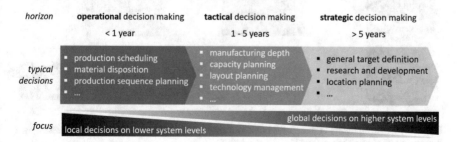

Fig. 2.10 Characteristics of strategic, tactical and operational planning, own illustration based on Dyckhoff and Spengler (2010)

2.3.2 Decision Theory

The most popular description of decision processes given by Simon describes decision making as a process with three consecutive phases (Simon 1977; Shim et al. 2002):

1. *intelligence phase* as the search for and description of problems
2. *design phase* as the development of alternatives
3. *choice phase* as the analysis and selection of alternatives for implementation

Similar models for decision making processes have been developed by Kepner and Tregoe (1976), Baker et al. (2002), Bazerman and Moore (2012). In the intelligence phase, the main challenge is to identify and describe the decision problem, taking into account organizational and personal objectives of the decision maker. With respect to the general problem structure, Simon describes a continuum from programmed (routine, repetitive, well structured, easy to solve) to non-programmed (new, novel, ill-structured, difficult to solve) (Simon 1977; Shim et al. 2002), whereby many practical problems are somewhere in between those two extremes (Turban et al. 2007). In the design phase, a decision model needs to be developed, abstracting reality but considering the relations between the most important variables. According to Laux (2005), such models base on a target function on the one hand and a decision field on the other hand (compare Fig. 2.11). Alternative solutions with their inherent sets of attributes need to be derived and criteria for their evaluation need to be defined, also taking into account surrounding conditions such as potential future developments. However, too many alternatives or attributes increase the complexity of decision making and may result in an information overload of the decision maker (Payne et al. 1992; Turban et al. 2007). By means of the decision model, the alternative options are evaluated and compared regarding their results. Making manufacturing decisions often necessitates to consider performance indicators from multiple target dimensions as introduced in Sect. 2.1.3 (Ferdows and De Meyer 1990).

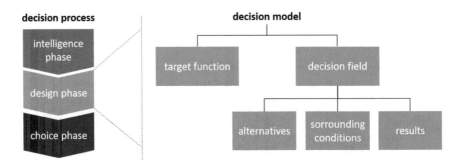

Fig. 2.11 Basic elements of decision models and corresponding phases of decision processes, own illustration based on Simon (1977) and Laux (2005)

In the choice phase, a selection must be made among the available alternatives. Here, one of the main challenges is to decide under which circumstances an alternative is preferred to another. Apart from the performance of each alternative, this may involve additional criteria such as risk-affinity, time issues or budget constraints. Following a normative selection principle, a decision maker will try to identify the best (i.e. optimal) solution (Turban et al. 2007). This assumption bases upon the theory of a *rational decision maker*,[1] who is able to rank the desirability of all possible consequences of an analysis (Rommelfanger and Eickemeier 2002; Turban et al. 2007). In particular for cases with a high complexity, the selection of alternatives should be carried out in a systematic manner, e.g. by applying methods for multi-criteria decision analysis (MCDA).

2.3.3 Multi-criteria Decision Analysis

As described in Sect. 2.1.3, decision making in manufacturing typically involves the consideration of multiple target dimensions. In the understanding of decision theory, this results in the need to evaluate several or all feasible alternative courses of action with their inherent attribute values regarding potential consequences for each target. The targets must be reflected by a set of meaningful criteria, i.e. performance indicators. Depending on the solution space, two general approaches can be distinguished (Zimmermann and Gutsche 1991; Belton and Steward 2002; Geldermann and Lerche 2014):

- **Multi-objective decision making (MODM)** methods are used to provide mathematically optimal solutions for problems with a continuous solution space, i.e. an infinite number of possible alternatives. Popular techniques are linear and non-linear programming, goal programming or (meta) heuristics.
- **Multi-attribute decision making (MADM)** methods are used to find good solutions for problems with a discrete solution space, i.e. a limited number of possible alternatives. Popular techniques include classical methods on the one hand such as analytical hierarchy process (AHP), analytical network process (ANP) and value benefit analyses (VBA) as well as outranking methods on the other hand. The latter group comprises but is not limited to methods like preference ranking organization method for enrichment of evaluations (PROMETHEE) and technique for order preference by similarity to ideal solution (TOPSIS).

Both approaches are highly relevant for decision making in the context of manufacturing, for instance in product design, supply chain design, production scheduling

---

[1]The *rational decision maker*, also known as *homo oeconomicus*, refers to a popular concept in economics, assuming humans as consistently rational and self-interested agents, who aim to maximize utility (consumer perspective) or profit (business perspective). The validity of the concept can be questioned, as real-world decision makers do not always act in a rational way. Discussions about this issue can be found in Schwartz (1972) and Halpern and Stern (1998). Nevertheless, the concept can still be regarded as useful when analyzing human behavior.

or end of life scenario selection. In particular, MCDA methods have become quite popular for decision making in cases where a balancing between economic and environmental issues is aspired (Zhou et al. 2013). Overviews about the application in the context of sustainable manufacturing are given by Ilgin and Gupta (2010) and Thies et al. (2019). MADM methods turned out to be more popular than MODM methods, as many manufacturing decisions are dealing with a limited set of distinct alternatives. In addition, it is often very challenging to formulate objective and constraint functions needed to apply MODM methods in industrial practice (Zhou et al. 2013).

2.3.4 Decision Support Systems

Decision making should be supported by systematic methods such as MCDA but also suitable (software) tools, often referred to as decision support systems (DSS). Several studies underline the high relevance of DSS for producing industries (Eom et al. 1998; Eom and Kim 2006; Kim and Eom 2016). As an example, Spengler et al. (1998) developed a DSS for the environmental assessment of recycling measures. An overview about various applications in manufacturing can be found in Kasie et al. (2017). Unfortunately, the terms decision support and DSS are not consistently used in literature. An overview about the evolution of DSS related terms and concepts can be found in Shim et al. (2002). Accordingly, DSS is often used interchangeably with the terms management support system (MSS) and business intelligence (BI) system. Turban et al. (2007) describe DSS as an "umbrella term", lacking a universally accepted definition. Thus, underlying technologies and capabilities differ significantly, comprising for instance data mining methods, operations research (OR) models and techniques, enterprise resource planning (ERP) systems, knowledge management systems (KMS), expert systems (ES) and knowledge-based systems (KBS) (Turban et al. 2007; Borissova and Mustakerov 2012). An early definition described DSS as "interactive computer-based systems, which help decision makers utilize data and models to solve unstructured problems" (Gorry and Morton 1971). A definition by Bonczek et al. (1980) understands DSS as computer-based systems, which consist of three interacting components: a language system as interface to the DSS user, a knowledge system containing domain specific knowledge and a problem-processing system with certain problem-solving capabilities. Kwon et al. (2005) provide an overview about different DSS types. Following their understanding, DSS are characterized by the existence of certain system elements such as database, model base and knowledge base as well as addressed user groups. Based on popular definitions of DSS, an ideal set of key characteristics and capabilities which typically characterize these systems is shown in Fig. 2.12.

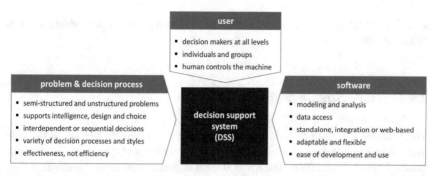

Fig. 2.12 Key characteristics and capabilities of decision support systems, own illustration based on Turban et al. (2007)

2.4 Modeling of Manufacturing Systems and Value Chains

As described in Sect. 2.1, manufacturing systems are complex systems with different elements that feature dynamic relations and interdependencies. In order to improve a manufacturing system's resource efficiency as a contribution to a sustainable manufacturing (compare Sect. 2.2), the use of system models is indispensable. Models are abstractions of real-world problems or systems, which can be used to experiment with alternative system configurations. They can help to judge about consequences of actions in advance, preventing from unfavorable (i.e. costly or disruptive) decisions (Law and Kelton 1991; Borshchev and Filippov 2004). Hence, models can allow decision makers to evaluate the consequences of their decisions before changes are implemented in the real-world system. In the context of a resource efficient manufacturing, well-established methods comprise but are not limited to value stream mapping (VSM), material and energy flow analysis (MEFA), material flow cost accounting (MFCA), life cycle assessment (LCA), life cycle costing (LCC), material flow simulation (MFS) and energy flow simulation (EFS). Figure 2.13 gives an

	scope				
	process	process chain	factory	value chain	life cycle
social	social life cycle assessment (SLCA)				
environmental	life cycle assessment (LCA)				
technical	material flow analysis (MFA), material & energy flow analysis (MEFA)				
	material flow simulation (MFS), energy flow simulation (EFS)				
	value stream mapping (VSM)				
economic	material flow cost accounting (MFCA)				
	life cycle costing (LCC)				

(evaluation dimensions)

Fig. 2.13 Methods for manufacturing system modeling and evaluation

overview about typical scopes and evaluation dimensions that are addressed with these methods.

2.4.1 Value Stream Mapping

Value stream mapping (VSM) is a methodology to systematically apply the lean manufacturing principle of waste reduction to a product's value stream (Rother and Shook 1999; Erlach 2010). The value stream is understood as the production flow from raw materials until final product. As indicated in Fig. 2.14, VSM tracks both material but also information flows along the value stream and makes them transparent using comprehensible visual representations, referred to as *value stream maps* (Rother and Shook 1999). Providing a common language, a value stream map serves as a good starting point to analyze the current state of a manufacturing system and reduce waste by applying lean principles (Erlach 2010). In this context, *waste* can refer to different types of resource wastage according to lean manufacturing principles like over-production, waiting, transport, over-processing, inventory, motion or defects (Ohno 1988; Womack and Jones 1996; Hines and Rich 1997). One basic principle of VSM is the orientation towards the actual customer demand for products. By synchronizing the production with the *customer takt time*, over-production and other wastage shall be reduced (Rother and Shook 1999; Erlach 2010). The intention of VSM is to improve the whole system instead of optimizing just parts. This means to follow the value stream across several companies in an ideal case in order to exploit the full potential of the method.

In recent years, various extensions of the traditional VSM approach have been developed. In energy value stream mapping (EVSM), an extension of material and information flows by energy flows broadens the scope of the methodology and allows to identify energy related wastages (Erlach and Westkämper 2009; Reinhart et al. 2010; Bogdanski et al. 2012; Posselt et al. 2014; Fischer et al. 2015).

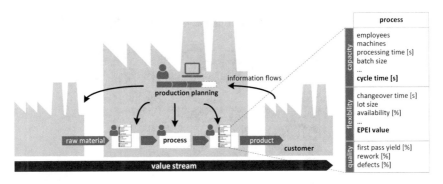

Fig. 2.14 Value stream mapping approach with standardized process description, adapted from Erlach (2010)

Quality value stream mapping (QVSM) puts special emphasis on the visualization of inspection processes, quality key indicators and quality control loops as well as quality related costs within the process chain (Haefner et al. 2014). Some researchers developed simulation-based VSM approaches in order to better analyze dynamics within manufacturing systems, which cannot be fully covered by the traditional static VSM approach (Sparks and Badurdeen 2014; Alvandi et al. 2016; Schöne-mann et al. 2016). Also, the limitation to single product systems is overcome by some of these approaches. Other authors integrated further environmental aspects leading to approaches such as sustainability value stream mapping or environmental value stream mapping (Torres and Gati 2009; Sproedt and Plehn 2012; Sparks and Badurdeen 2014; Faulkner and Badurdeen 2014).

2.4.2 Material and Energy Flow Analysis

Material flow analysis (MFA) describes a systematic assessment of material flows and stocks in a defined system (Brunner and Rechberger 2016). The method uses material balances indicating sources, pathways, intermediate and final sinks for materials. Going back to the input–output analysis introduced by Leontief in the 1930s, the application of MFA as a distinct methodology has evolved from the 1970s on Suh (2005) and Brunner and Rechberger (2016). Material and energy flow analysis (MEFA) extends the scope of the methodology to energy flows. Due to its simplicity, MFA/MEFA has become well-established in the context of economics, resource management, waste management and environmental management. Examples in the context of resource efficient manufacturing can for instance be found in Haberl et al. (2004), Ghadimi et al. (2014), Lambrecht and Thißen (2015), Thiede et al. (2016), Blume et al. (2017). Notations used in the MEFA method are not generally agreed but vary from study to study. A comprehensible notation by Suh is presented in Fig. 2.15, based on the work of Hannon (1973) and Finn (1976). MEFA can also be regarded as method to calculate the life cycle inventory (LCI) for a life cycle assessment (LCA) study (Brunner and Rechberger 2016). The other way around, life cycle impact assessment (LCIA) as part of the LCA methodology can be applied to MEFA results in order to receive transparency about environmental impacts related to the mapped material and energy flows (compare Sect. 2.4.5). A main difference between MEFA and LCA exists in the level of modeling detail: while LCA strives to assess all flows with a potential relevance in terms of environmental impacts, MEFA typically tends to reduce the number of considered flows in order to maintain a better manageability (Brunner and Rechberger 2016).

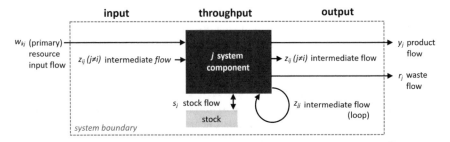

Fig. 2.15 MEFA based flow diagram for a generalized input–output system, adapted from Suh (2005)

In Fig. 2.15, the following notations are used:

i, j	system components
w_{kj}	resource input flows from a source k outside the system boundaries to j
z_{ij}, z_{ji}	intermediate flows between j and another system element i
z_{jj}	intermediate flows looping from j to j
r_j	waste flows from j leaving the system
y_j	usable flows (product flows) from j leaving the system
s_j	stock changes of j due to imbalances in entering and leaving flows

2.4.3 Material Flow Cost Accounting

The method material flow cost accounting (MFCA) builds upon the MEFA method, calculating related costs for energy and material balances within a defined system (compare Fig. 2.16). It is specifically designed for manufacturing industries (Jasch 2009). The method has reached a high popularity in particular in Japan and Germany since the 1990s (Fichter et al. 1997; Nakajima 2009) and a general framework for MFCA was established in 2011 with the standard ISO 14051 (International Organization for Standardization 2011). According to this framework, the method puts a special focus on the comparison of costs related with products and material losses i.e. waste flows. This can lead to fundamental changes in a company's way of decision

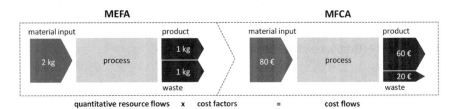

Fig. 2.16 Exemplary MFCA results for a machining process with underlying MEFA model, visualized as Sankey diagram

making, as waste-induced inefficiencies becomes transparent compared to traditional accounting (Jasch 2009; Viere et al. 2011). The scope of an MFCA analysis is not limited to single companies, but can be extended to value chains in order to improve material and energy efficiency on a greater scale (Jasch 2009; International Organization for Standardization 2011). In this way, the method also intends to reduce environmental impacts, although an environmental evaluation is not carried out explicitly (Fichter et al. 1997; International Organization for Standardization 2011; Viere et al. 2011). Thus, MFCA can also be seen as a starting point for more comprehensible environmental assessments such as LCA in order to combine economic and environmental analyses (Viere et al. 2011).

2.4.4 Life Cycle Costing

Life cycle costing (LCC) is a method used for the assessment of capital investment options under consideration of life cycle costs, i.e. expected future costs and revenues (Verein Deutscher Ingenieure 2005; Götze 2010). It extends the scope of an MFCA by other aspects than material and energy flows like initial invests, maintenance or disposal costs. The motivation to conduct LCC evolves from the observation, that a significant share of costs related to a product is due to resource consumption and maintenance costs in its use stage (Denkena et al. 2005; Abele et al. 2009; Noske and Kalogerakis 2009; Lauven et al. 2010). Figure 2.17 presents a general example, where costs during and after utilization clearly exceed the costs caused by development and manufacturing of a good. In contrast, the ability to influence the life cycle costs is highest in the early phases (Schild 2005). Therefore, LCC takes into account the product life cycle from the perspective of the product user, i.e. costs related with acquisition, use and disposal (Abele et al. 2009). LCC methodologies have been anchored in different standards such as VDI 2884, VDMA 34160 and DIN EN 60300 (Verein Deutscher Ingenieure 2005; Verband Deutscher Maschinen- und Anlagenbau e.V. 2007; Deutsches Institut für Normung 2014). Depending on

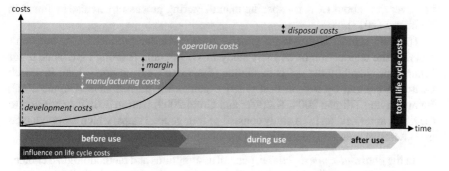

Fig. 2.17 Graphical representation of life cycle costs, adapted from Verein Deutscher Ingenieure (2005)

the framework, definitions of life cycle stages, accountable costs or calculation rules differ (Bünting 2009). In this context, the total costs of ownership (TCO) are often used as equivalent to LCC. As an example, typical life cycle costs of energy related production equipment can be found in Müller et al. (2009), comprising e.g. costs for piping systems and electrical installations, heat demands, electrical energy demands, replacement of filters, leakage repair or disposal of components. The actual calculation of life cycle costs applies typical methods of investment accounting like the net present value method (Müller et al. 2009; Götze 2010). Although LCC is a useful method to take investment decisions, not only economic aspects should be taken into account for decision making. For an evaluation of environmental aspects, a life cycle assessment (LCA) study can be conducted (compare Sect. 2.4.5). In order to estimate trade-offs between economic, environmental and technical aspects over the life cycle of an investment good, the method of life cycle performance evaluation can be applied (Niggeschmidt et al. 2013).

2.4.5 Life Cycle Assessment

Life cycle assessment (LCA) basically refers to a methodology for the environmental evaluation of products. According to ISO standards 14040 and 14044, LCA can be used as decision support tool with the goals to improve the environmental performance of products, inform decision makers, select relevant indicators for environmental performance and support marketing activities (International Organization for Standardization 2006a, b). It considers the whole product life cycle including raw material extraction, production, use and end of life treatment (cradle to grave) and can thus help to avoid problem shifting along different life cycle stages. Apart from products, LCA can also be applied to organizations, processes, consumers or countries (Herrmann 2010; Hellweg and Mila i Canals 2014). An explicit methodological adaption for manufacturing systems is for instance presented by Löfgren et al. (2011). LCA for a wide range of single manufacturing (unit) processes have been compiled in the *CO₂PE!* project (Kellens et al. 2012; Kellens 2013). A literature overview about LCA for specific manufacturing processes can also be found in Reinhardt (2013).

The ISO standards provide a general framework for product LCA, but don't describe the conduction of an LCA study in very detail. Thorough guidance is for instance provided by the ILCD handbook (European Commission – Joint Research Centre 2010) and by several best practice guides for practitioners (Guinée 2002; Baumann and Tillman 2004; Klöpffer and Grahl 2009; Curran 2012). According to the ISO framework, an LCA study consists of four general steps, which are conducted in an iterative manner (International Organization for Standardization 2006a, b):

1. In the *goal and scope definition*, general assumptions and methodological choices need to be made. In this context, the definition of a *functional unit* is required, serving as measurable function to allow for a comparison of different products

or processes (e.g. packaging for one liter of water). Further, *system boundaries* need to be defined in order to determine the elements and flows to consider in the study.

2. In the *life cycle inventory (LCI)* analysis, resource flows within the system boundaries need to be quantified. With respect to manufacturing systems, this means for instance to quantify the direct input and output flows in terms of energy and materials for manufacturing processes according to the MEFA approach (compare Sect. 2.4.2). Quantification ideally refers to direct measurements, but can also be based on calculations or well-founded estimations (Klöpffer and Grahl 2009). In practical LCA application, it is often impossible to collect primary data for all resource flows needed to manufacture a product. In this case, LCA databases can be used as source for LCI data. They contain aggregated datasets representing the production of various materials, (pre)products and energies (United Nations Environment Programme 2011). Well-established databases for product and process related LCA data are for instance the *Life Cycle Data Network (LCDN)* and the *ELCD database* (European Commission – Joint Research Centre 2019a, b) as well as commercial databases provided by *ecoinvent* (Frischknecht and Rebitzer 2005; Ecoinvent Centre – Swiss Centre for Life Cycle Inventories 2018) and *Thinkstep* (Thinkstep AG 2018). A good overview about available databases and in particular access to open source datasets is provided by the web portal *openLCA Nexus* (GreenDelta GmbH 2018).

3. The *life cycle impact assessment (LCIA)* focuses on the evaluation of potential environmental impacts related to the flows quantified in the LCI. A wide range of available methods exists to classify and characterize environmental impacts in an LCA study. Popular LCIA methods comprise for example *Eco-Indicator* (Goedkoop and Spriensma 2001), *CML* (Guinée 2002) or *ReCiPe* method (Goedkoop et al. 2009). The methods differ regarding used equivalency factors (i.e. the assumed significance of an element's environmental impacts, also referred to as characterization factors) and the impact categories assessed (Guinée 2002; Curran 2012). As shown in Fig. 2.18, the evaluation can be either focusing on midpoint impacts (e.g. climate change, ozone depletion, human toxicity) or end point impacts (e.g. human health or natural environment). Endpoint impacts are obtained by normalization and weighting of midpoint related impacts. Some methods such as *Eco-Indicator 99* allow to condense all impacts to a single score by further weighting, facilitating a quicker comparison of competing products (Goedkoop and Spriensma 2001).

4. In the *interpretation phase* the results are critically discussed and conclusions or recommendations for further actions are derived. This phase may also involve an iterative process of reviewing and revising goal and scope as well as collected data (International Organization for Standardization 2006b). Further, sensitivity analyses may be conducted in order to validate the results (Guinée 2002).

If complex product systems and different impact categories are investigated, specific software is typically used to conduct LCA studies. A systematic comparison for some of the most popular commercial and open source software solutions can

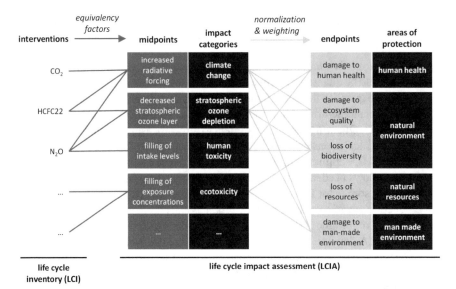

Fig. 2.18 Distinction between midpoint and endpoint impacts in LCA with related chains of effects, adapted from Guinée (2002)

be found in Lüdemann and Feig (2014). In practice, the tools may also be used to create simplified *screening LCA*, which do not comply with the requirements for a full LCA according to ISO framework (Guinée 2002; International Organization for Standardization 2006b).

The challenges related with an environmental evaluation of products over their life cycles are highly case specific. Thus, methodological developments are continuously ongoing (Finnveden et al. 2009; Hellweg and Mila i Canals 2014). The static character of LCA is an often criticized aspect, which impedes the application of LCA as decision making tool (Löfgren and Tillman 2011). The low resolution does not allow for an assessment of dynamic interrelationships of manufacturing processes. Thus, a coupling of LCA with other tools such as simulation (compare Sect. 2.4.6) is an often discussed solution to overcome these shortcomings (Sproedt et al. 2015; Stiel et al. 2016). Alternative manufacturing scenarios can then be better compared regarding their environmental impacts from a life cycle perspective (Widok et al. 2012). Further, an integration with real-time production information is discussed (Cerdas et al. 2017), as commercial software tools do not fully provide such functionalities so far (Löfgren and Tillman 2011; Thiede et al. 2013; Stiel et al. 2016). Due to the significant efforts that go along with the conduction of an LCA study as well as the great influence of underlying assumptions, *integrated computational (IC)* methods are proposed to ease the method's applicability and to improve its use value. Software tools providing the IC approach can be used to assess great numbers of differing *what if* scenarios at once in a highly automated manner (Cerdas et al. 2018). Regionalized LCA constitute another example to tackle uncertainties in LCA studies due to underlying assumptions

(Hellweg and Mila i Canals 2014). By taking into account regional effects and site-specific production conditions instead of using global average data, differing regional recommendations regarding the production or usage of products can be provided e.g. for electric vehicle use (Egede 2017) or fresh water consumption (Boulay et al. 2011). Social life cycle assessment (SLCA) as further methodological extension of LCA aims at an integration of social aspects (Grießhammer et al. 2006; Dreyer et al. 2010). In recent years, a *social hotspot database* has been launched, containing information about social consequences of production along value chains (Social Hotspot Database 2019). Further, the *UNEP/SETAC Life Cycle Initiative* developed guidelines for the assessment of social impacts of products (United Nations Environment Programme 2009). The increasing importance of social aspects is also expressed by the development of the ISO 26000 standard, providing guidance for organizations in terms of social responsibility (International Organization for Standardization 2010). However, a comprehensible method to assess all dimensions of sustainability in an integrated LCA approach is not yet established (Hellweg and Mila i Canals 2014).

2.4.6 Material and Energy Flow Simulation

In contrast to the rather static methods presented up to here, simulation allows to investigate time-dependent events and interactions within a system. It can be understood as "imitation of the operation of a real-world process or system over time" (Banks 1998). The method uses dynamic system models to describe and analyze the behavior of existing or planned systems with the goal to transfer the results from model to real-world (Banks 1998; Verein Deutscher Ingenieure 2018). Typical procedures for conducting simulation studies can for instance be found in Banks et al. (2010) and Stoldt and Putz (2017). The term simulation encompasses different simulation paradigms such as discrete event simulation (DES), agent based simulation (ABS), dynamic systems (DS) simulation and system dynamics (SD) simulation (Borshchev and Filippov 2004). The paradigms differ with respect to their level of abstraction compared with the real-world system. Further, a distinction is made between discrete and continuous systems or system models, respectively. For the analysis of material flows in manufacturing systems, typically the DES paradigm is applied (Klingstam and Gullander 1999). Material flow simulation (MFS) depicts the time-dependent movement of materials along a process chain or value chain (European Commission 2014). MFS traditionally focuses on the analysis of manufacturing system performance in terms of technical targets like output quantities, lead times or stocks with the intention to increase profitability (Putz et al. 2011; Verein Deutscher Ingenieure 2018). Typical elements to be used in MFS are moving objects like products and transport equipment, stationary elements such as machines or material buffers and organizational elements like break times, machine failures or shift plans (Junge 2007). Energy flow simulation (EFS) extends the flows considered in an MFS model by energy flows, usually building on the existence of a DES model representing the material flow. Discussions, examples and procedure models for the

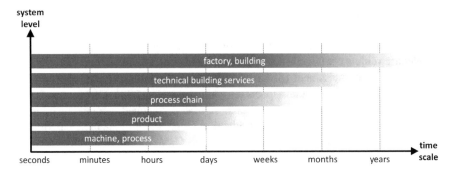

Fig. 2.19 Spatial and temporal scales to consider for production system simulation, adapted from Schönemann (2017)

combination of material and energy flows in simulation can be found in Thiede (2012) and Stoldt and Putz (2017). In addition, MFS and EFS can be coupled with the simulation of other system elements of factories, for example building shell or human worker models (Stahl et al. 2013; Schönemann 2017). Challenges arise due to the different time scales that are typically addressed in dependence on modeled system elements and manufacturing system levels (compare Fig. 2.19). In practical application, specific simulation software is used to build up simulation models and analyze their behavior (Banks et al. 2010; Verein Deutscher Ingenieure 2018). Popular tools for MFS and EFS are for instance *MATLAB, Vensim, Plant Simulation, Quest* or *Anylogic*.

Simulation is broadly applied in the context of manufacturing. An overview about the application of DES for manufacturing system design and operation is provided by Negahban and Smith (2014). The authors state that simulation is increasingly described as tool for decision making and the interest of applying simulation in industrial contexts is constantly growing. The European Union regards simulation and forecasting technologies (S&FT) as a main lever to increase efficiency and sustainability of manufacturing systems. Therefore, a joint European strategy for further developing these technologies has been elaborated in the *Pathfinder* project (European Commission 2014). Accordingly, emerging S&FT comprise for instance the development of integrated multi-disciplinary and multi-domain technologies (e.g. tools for sustainable supply chain design), life cycle management tools (e.g. for end of life processes) and multi-level simulation (e.g. enhancing data integration across process chains).

References

Abele E, Dervisopoulos M, Kuhrke B (2009) Bedeutung und Anwendung von Lebenszyklusanalysen bei Werkzeugmaschinen. Lebenszykluskosten optimieren. Gabler, Wiesbaden, pp 51–80

Abreu MF, Alves AC, Moreira F (2017) Lean-Green models for eco-efficient and sustainable production. Energy 137:846–853. https://doi.org/10.1016/j.energy.2017.04.016

Alvandi S, Li W, Schönemann M et al (2016) Economic and environmental value stream map (E2VSM) simulation for multi-product manufacturing systems. Int J Sustain Eng 9(6):354–362. https://doi.org/10.1080/19397038.2016.1161095

Baker D, Bridges D, Johnson RH et al (2002) Guidebook to decision-making methods. US Department of Energy, Washington

Banks J (1998) Handbook of simulation: principles, methodology, advances, applications, and practice. Wiley, Atlanta, Georgia

Banks J, Carson JS, Nelson BL, Nicol D (2010) Discrete-event system simulation. Pearson, Upper Saddle River

Baumann H, Tillman A-M (2004) The Hitch Hiker's guide to LCA. Studentlitteratur, Lund

Bazerman MH, Moore DA (2012) Judgement in managerial decision making. Wiley, Hoboken

Beer S (1995) Diagnosing the system for organizations. Wiley, Chichester

Belton V, Steward T (2002) Multiple criteria decision analysis – An integrated approach. Springer US

Blume S, Kurle D, Herrmann C, Thiede S (2017) Toolbox for increasing resource efficiency in the European metal mechanic sector. Procedia CIRP 61:40–45. https://doi.org/10.1016/j.procir.2016.11.247

Bogdanski G, Schönemann M, Thiede S, et al (2012) An extended energy value stream approach applied on the electronics industry. In: IFIP international conference on advances in production management systems. Springer, pp 65–72

Bonczek RH, Holsapple CW, Whinston AB (1980) The evolving roles of models in Decision Support Systems. Decis Sci 11(2):337–356

Borissova D, Mustakerov I (2012) An integrated framework of designing a decision support system for engineering predictive maintenance. Int J Inf Technol Knowl 6(4):366–376

Borshchev A, Filippov A (2004) From system dynamics and discrete event to practical agent based modeling: reasons, techniques, tools. In: The 22nd International conference of the system dynamics society, 25–29 July 2004. Oxford, Oxford, England

Botta V (1997) Kennzahlensysteme als Führungsinstrumente. Erich Schmidt Verlag, Berlin

Boulay A-M, Bulle C, Bayart J-B et al (2011) Regional Characterization of freshwater use in LCA: modeling direct impacts on human health. Environ Sci Technol 45(20):8948–8957. https://doi.org/10.1021/es1030883

Braungart M, McDonough W, Bollinger A (2007) Cradle-to-cradle design: creating healthy emissions – A strategy for eco-effective product and system design. J Clean Prod 15(13–14):1337–1348. https://doi.org/10.1016/j.jclepro.2006.08.003

Brown MG (1997) Kennzahlen - Harte und weiche Faktoren erkennen, messen und bewerten. Hanser, München

Brunner PH, Rechberger H (2016) Practical handbook of material flow analysis: for environmental, resource, and waste engineers. CRC Press, Boca Raton

Bunse K, Vodicka M, Schönsleben P et al (2011) Integrating energy efficiency performance in production management – gap analysis between industrial needs and scientific literature. J Clean Prod 19(6–7):667–679. https://doi.org/10.1016/j.jclepro.2010.11.011

Bünting F (2009) Lebenszykluskostenbetrachtungen bei Investitionsgütern. Lebenszykluskosten optimieren. Gabler, Wiesbaden, pp 35–50

Cantner U, Krüger J, Hanusch H (2007) Produktivitäts- und Effizienzanalyse. Der nicht-parametrische Ansatz. Springer, Berlin, Heidelberg, New York

Cerdas F, Thiede S, Herrmann C (2018) Integrated computational life cycle engineering – application to the case of electric vehicles. CIRP Ann 67(1):25–28. https://doi.org/10.1016/j.cirp.2018.04.052

Cerdas F, Thiede S, Juraschek M et al (2017) Shop-floor life cycle assessment. Procedia CIRP 61:393–398. https://doi.org/10.1016/j.procir.2016.11.178

Chryssolouris G (2006) Manufacturing systems: theory and practice, 2nd edn. Springer, New York

Curran MA (2012) Life cycle assessment handbook. Scrivener Publishing LCC, Salem

Dehning P, Thiede S, Mennenga M, Herrmann C (2017) Factors influencing the energy intensity of automotive manufacturing plants. J Clean Prod 142:2305–2314. https://doi.org/10.1016/j.jcl epro.2016.11.046

Denkena B, Harms A, Vogeler S, Noske H (2005) Können teure Werkzeugmaschinen auf längere Sicht günstiger sein? Werkstatttechnik Online 95:519–523

Despeisse M (2013) Sustainable manufacturing tactics and improvement methodology: a structured and systematic approach to identify improvement opportunities. Cranfield University

Despeisse M, Ball P, Evans S (2013) Strategies and ecosystem view for industrial sustainability. In: 20th CIRP international conference on life cycle engineering, pp 565–570. https://doi.org/10.1007/978-981-4451-48-2

Despeisse M, Ball PD, Evans S, Levers A (2012) Industrial ecology at factory level: a prototype methodology. Proc Inst Mech Eng Part B J Eng Manuf 226(10):1648–1664. https://doi.org/10.1177/0954405412449937

Deutsches Institut für Normung (2003) DIN 8580:2003–09 - Fertigungsverfahren - Begriffe, Einteilung

Deutsches Institut für Normung (2014) DIN EN 60300–3–3: Zuverlässigkeitsmanagement - Teil 3–3: Anwendungsleitfaden - Lebenszykluskosten

Diaz-Elsayed N, Jondral A, Greinacher S et al (2013) Assessment of lean and green strategies by simulation of manufacturing systems in discrete production environments. CIRP Ann 62(1):475–478. https://doi.org/10.1016/j.cirp.2013.03.066

Diekmann J, Eichhammer W, Neubert A et al (1999) Energie-Effizienz-Indikatoren. Springer, Berlin, Heidelberg

Dreyer LC, Hauschild MZ, Schierbeck J (2010) Characterisation of social impacts in LCA: Part 1: development of indicators for labour rights. Int J Life Cycle Assess 15(3):247–259. https://doi.org/10.1007/s11367-009-0148-7

Dües CM, Tan KH, Lim M (2013) Green as the new Lean: how to use Lean practices as a catalyst to greening your supply chain. J Clean Prod 40:93–100. https://doi.org/10.1016/j.jclepro.2011.12.023

Duflou JR, Sutherland JW, Dornfeld D et al (2012) Towards energy and resource efficient manufacturing: a processes and systems approach. CIRP Ann Manuf Technol 61(2):587–609. https://doi.org/10.1016/j.cirp.2012.05.002

Dyckhoff H (2006) Produktionstheorie - Grundzüge industrieller Produktionswirtschaft, 5th edn. Springer, Berlin, Heidelberg, New York

Dyckhoff H, Souren R (2007) Nachhaltige Unternehmensführung: Grundzüge industriellen Umweltmanagements. Springer, Berlin, Heidelberg

Dyckhoff H, Spengler T (2010) Produktionswirtschaft, 3rd edn. Springer, Heidelberg, Dordrecht, London, New York

Ecoinvent Centre – Swiss Centre for Life Cycle Inventories (2018) Ecoinvent database v3.5. https://www.ecoinvent.org/. Accessed 9 Aug 2019

Egede P (2017) Environmental assessment of lightweight electric vehicles. Springer, Cham

Eom S, Kim E (2006) A survey of decision support system applications (1995–2001). J Oper Res Soc 57(11):1264–1278

Eom S, Lee S, Kim E, Somarajan C (1998) A survey of decision support system applications (1988–1994). J Oper Res Soc 49:109–120

Erlach K (2010) Wertstromdesign. Springer, Berlin, Heidelberg

Erlach K, Westkämper E (2009) Energiewertstrom: Der Weg zur energieeffizienten Fabrik. Fraunhofer Verlag, Stuttgart

European Commission (1975) Waste framework directive 75/442/EEC

European Commission (2014) Pathfinder – White-paper

European Commission – Joint Research Centre (2016) Production costs from energy-intensive industries in the EU and third countries. https://publications.jrc.ec.europa.eu/repository/bitstr eam/JRC100101/ldna27729enn.pdf

European Commission – Joint Research Centre (2010) International reference life cycle data system (ILCD) handbook – general guide for life cycle assessment – Detailed guidance. https:// lct.jrc.ec.europa.eu/pdf-directory/ILCD-Handbook-General-guide-for-LCA-DETAIL-online-12March2010.pdf

European Commission – Joint Research Centre (2019a) ELCD database. https://eplca.jrc.ec.eur opa.eu/ELCD3/. Accessed 10 Aug 2019

European Commission – Joint Research Centre (2019b) The life cycle data network. https://eplca. jrc.ec.europa.eu/LCDN/. Accessed 13 Aug 2019

Fantini P, Palasciano C, Taisch M (2015) Back to Intuition: proposal for a performance indicators framework to facilitate eco-factories management and benchmarking. Procedia CIRP 26:1–6. https://doi.org/10.1016/j.procir.2014.07.099

Faulkner W, Badurdeen F (2014) Sustainable value stream mapping (Sus-VSM): methodology to visualize and assess manufacturing sustainability performance. J Clean Prod 85:8–18. https://doi. org/10.1016/j.jclepro.2014.05.042

Ferdows K, De Meyer A (1990) Lasting improvements in manufacturing performance: in search of a new theory. J Oper Manag 9(2):168–184. https://doi.org/10.1016/0272-6963(90)90094-T

Fichter K, Loew T, Seidel E (1997) Betriebliche Umweltkostenrechnung - Methoden und praxisgerechte Weiterentwicklung. Springer, Berlin, Heidelberg

Finn J (1976) Measures of structure and functioning derived from analysis of flows. J Theor Biol 56(23):363–380

Finnveden G, Hauschild MZ, Ekvall T et al (2009) Recent developments in life cycle assessment. J Environ Manage 91(1):1–21. https://doi.org/10.1016/j.jenvman.2009.06.018

Fischer J (2017) Prioritizing Components for Lean and Green Manufacturing. Vulkan, Essen

Fischer J, Weinert N, Herrmann C (2015) Method for selecting improvement measures for discrete production environments using an extended energy value stream model. Procedia CIRP 26:133–138. https://doi.org/10.1016/j.procir.2014.07.100

Frischknecht R, Rebitzer G (2005) The ecoinvent database system: a comprehensive web-based LCA database. J Clean Prod 13(13–14):1337–1343. https://doi.org/10.1016/j.jclepro.2005.05.002

Geldermann J, Lerche N (2014) Leitfaden zur Anwendung von Methoden der multikriteriellen Entscheidungsunterstützung. Georg-August-Universität Göttingen

Ghadimi P, Li W, Kara S, Herrmann C (2014) Integrated material and energy flow analysis towards energy efficient manufacturing. Procedia CIRP 15:117–122. https://doi.org/10.1016/j. procir.2014.06.010

Goedkoop M, Heijungs R, Huijbregts M et al (2009) ReCiPe 2008 – A life cycle impact assessment method which comprises harmonised category indicators at the midpoint and the endpoint level. Ministerie van VROM, Den Haag

Goedkoop M, Spriensma R (2001) The Eco-indicator 99 – A damage oriented method for life cycle impact assessment – Methodology report. Pre Consultants, Amersfoort

Gorry GA, Morton MSS (1971) A framework for management information systems. Massachusetts Institute of Technology, Cambridge

Götze U (2010) Kostenrechnung und Kostenmanagement. Springer, Berlin, Heidelberg, New York

GreenDelta GmbH (2018) openLCA Nexus. https://nexus.openlca.org/. Accessed 10 Aug 2019

Greinacher S, Moser E, Hermann H, Lanza G (2015) Simulation based assessment of lean and green strategies in manufacturing systems. Procedia CIRP 29:86–91. https://doi.org/10.1016/j.procir. 2015.02.053

Grießhammer R, Benoît C, Dreyer LC et al (2006) Feasibility study: integration of social aspects into LCA. Eco-Institute, Freiburg

Größler A, Grübner A (2006) An empirical model of the relationships between manufacturing capabilities. Int J Oper Prod Manag 26(5):458–485. https://doi.org/10.1108/01443570610659865

Guinée JB (2002) Handbook on life cycle assessment – operational guide to the ISO standards. Kluwer Academic Publishers, Dordrecht

Günther E (1994) Ökologieorientiertes Controlling: Konzeption eines Systems zur ökologieorientierten Steuerung und empirische Validierung. Vahlen, München

Günther H-O, Tempelmeier H (2005) Produktion und Logistik. Springer, Berlin, Heidelberg, New York

Haag H (2013) Eine Methodik zur modellbasierten Planung und Bewertung der Energieeffizienz in der Produktion. Fraunhofer Verlag, Stuttgart

Haberl H, Fischer-Kowalski M, Krausmann F et al (2004) Progress towards sustainability? What the conceptual framework of material and energy flow accounting (MEFA) can offer. Land Use Policy 21(3):199–213. https://doi.org/10.1016/j.landusepol.2003.10.013

Haefner B, Kraemer A, Stauss T, Lanza G (2014) Quality Value Stream Mapping. Procedia CIRP 17:254–259. https://doi.org/10.1016/j.procir.2014.01.093

Halpern JJ, Stern RN (1998) Debating rationality: nonrational aspects of organizational decision making. Cornell University Press, New York

Hannon B (1973) The structure of ecosystems. J Theor Biol 41(3):535–546

Hansen W, Christopher M, Verbuecheln M (2002) EU Waste policy and challenges for regional and local authorities. Ecologic, Institute for International and European Environmental Policy, Berlin

Hauschild MZ (2015) Better – But is it good enough? on the need to consider both eco-efficiency and eco-effectiveness to gauge industrial sustainability. Procedia CIRP 29:1–7. https://doi.org/10.1016/j.procir.2015.02.126

Hauschild MZ, Herrmann C, Kara S (2017) An Integrated framework for life cycle engineering. Procedia CIRP 61:2–9. https://doi.org/10.1016/j.procir.2016.11.257

Heinemann T (2016) Energy and resource efficiency in aluminium die casting. Springer, Cham

Hellweg S, Mila i Canals L (2014) Emerging approaches, challenges and opportunities in life cycle assessment. Science (80-) 344(6188):1109–1113. https://doi.org/10.1126/science.1248361

Helu M, Rühl J, Dornfeld D, et al (2011) Evaluating trade-offs between sustainability, performance, and cost of green machining technologies. In: Glocalized solutions for sustainability in manufacturing: proceedings of the 18th CIRP international conference on life cycle engineering, pp 195–200

Herrmann C (2010) Ganzheitliches Life Cycle Management. Springer, Heidelberg, Dordrecht, London, New York

Herrmann C, Bergmann L, Thiede S, Halubek P (2007) Total life cycle management – An integrated approach towards sustainability. 3rd International conference on life cycle management

Herrmann C, Schmidt C, Kurle D et al (2014) Sustainability in manufacturing and factories of the future. Int J Precis Eng Manuf Green Technol 1(4):283–292. https://doi.org/10.1007/s40684-014-0034-z

Herrmann C, Thiede S, Heinemann T (2010) Ganzheitliche Ansätze zur Erhöhung der Energie- und Ressourceneffizienz in der Produktion. 10 Karlsruher Arbeitsgespräche Produktionsforsch 2010

Herrmann C, Thiede S, Stehr J, Bergmann L (2008) An environmental perspective on lean production. Manuf Syst Technol New Front 83–88. https://doi.org/10.1007/978-1-84800-267-8_16

Hesselbach J (2012) Energie- und klimaeffiziente Produktion. Vieweg+Teubner Verlag, Wiesbaden

Hines P, Rich N (1997) The seven value stream mapping tools. Int J Oper Prod Manag 17(1):46–64. https://doi.org/10.1108/01443579710157989

Huber J (1994) Nachhaltige Entwicklung durch Suffizienz, Effizienz und Konsistenz. In: Fritz P, Huber J, Levi H (eds) Nachhaltigkeit in naturwissenschaftlicher und sozialwissenschaftlicher Perspektive. Hirzel Wissenschaftliche Verlagsgesellschaft, Stuttgart, pp 31–46

Ilgin MA, Gupta SM (2010) Environmentally conscious manufacturing and product recovery (ECMPRO): a review of the state of the art. J Environ Manage 91(3):563–591. https://doi.org/10.1016/j.jenvman.2009.09.037

Institution of Mechanical Engineers (2009) The energy hierarchy. London

International Organization for Standardization (2013) DIN EN ISO 14031 – Umweltmanagement – Umweltleistungsbewertung

International Organization for Standardization (2011) DIN EN ISO 14051 – Umweltmanagement – Materialflusskostenrechnung - Allgemeine Rahmenbedingungen

International Organization for Standardization (2006a) ISO 14044 – Environmental management – Life cycle assessment – Requirements and guidelines

International Organization for Standardization (2006b) ISO 14040 – Environmental management – Life Cycle Assessment – Principles and Framework

International Organization for Standardization (2010) DIN ISO 26000 – Leitfaden zur gesellschaftlichen Verantwortung

Jasch C (2009) Environmental and material flow cost accounting. Springer, Dordrecht

Junge M (2007) Simulationsgestützte Entwicklung und Optimierung einer energieeffizienten Produktionssteuerung. Kassel University Press, Kassel

Kasie FM, Bright G, Walker A (2017) Decision support systems in manufacturing: a survey and future trends. J Model Manag 12(3):432–454. https://doi.org/10.1108/JM2-02-2016-0015

Kellens K (2013) Energy and resource Efficient Manufacturing – Unit process analysis and optimisation. KU Leuven

Kellens K, Dewulf W, Overcash M, et al (2012) Methodology for systematic analysis and improvement of manufacturing unit process life cycle inventory (UPLCI) – CO2PE! initiative (cooperative effort on process emissions in manufacturing). Part 1: Methodology description. Int J Life Cycle Assess 17(2):69–78. https://doi.org/10.1007/s11367-011-0352-0

Kepner CH, Tregoe BB (1976) The rational manager: a systematic approach to problem solving and decision making. Kepner-Tregoe Inc., Princeton

Kim EB, Eom SB (2016) Decision support systems application development trends (2002–2012). Int J Inf Syst Serv Sect 8(2):1–13. https://doi.org/10.4018/IJISSS.2016040101

Klingstam P, Gullander P (1999) Overview of simulation tools for computer-aided production engineering. Comput Ind 38(2):173–186. https://doi.org/10.1016/S0166-3615(98)00117-1

Klöpffer W, Grahl B (2009) Life cycle assessment – A guide to best practice. Wiley-VCH, Weinheim

Kwon O, Yoo K, Suh E (2005) UbiDSS: A proactive intelligent decision support system as an expert system deploying ubiquitous computing technologies. Expert Syst Appl 28(1):149–161. https://doi.org/10.1016/j.eswa.2004.08.007

Lambrecht H, Thißen N (2015) Enhancing sustainable production by the combined use of material flow analysis and mathematical programming. J Clean Prod 105:263–274. https://doi.org/10.1016/j.jclepro.2014.07.053

Lanza G, Ude J (2010) Multidimensional evaluation of value added networks. CIRP Ann Manuf Technol 59(1):489–492. https://doi.org/10.1016/j.cirp.2010.03.080

Lauven L, Wiedenmann S, Geldermann J (2010) Lebenszykluskosten als Entscheidungshilfe beim Erwerb von Werkzeugmaschinen. Georg-August-Universität Göttingen

Laux H (2005) Entscheidungstheorie. Springer, Berlin, Heidelberg, New York

Law AM, Kelton WD (1991) Simulation modeling and analysis. McGraw-Hill, New York

Löfgren B, Tillman AM (2011) Relating manufacturing system configuration to life-cycle environmental performance: Discrete-event simulation supplemented with LCA. J Clean Prod 19(17–18):2015–2024. https://doi.org/10.1016/j.jclepro.2011.07.014

Löfgren B, Tillman AM, Rinde B (2011) Manufacturing actor's LCA. J Clean Prod 19(17–18):2025–2033. https://doi.org/10.1016/j.jclepro.2011.07.008

Lüdemann L, Feig K (2014) Comparison of software solutions for Life Cycle Assessment (LCA) – A software ergonomic analysis. Logist J. https://doi.org/10.2195/lj_NotRev_luedemann_de_201409_01

Macharzina K, Wolf J (2008) Unternehmensführung: das internationale Managementwissen; Konzepte - Methoden – Praxis. Springer, Wiesbaden

Mentzer JT, DeWitt W, Keebler JS et al (2001) Defining supply chain management. J Bus Logist 22(2):1–25. https://doi.org/10.1002/j.2158-1592.2001.tb00001.x

Muchiri P, Pintelon L (2008) Performance measurement using overall equipment effectiveness (OEE): literature review and practical application discussion. Int J Prod Res 46(13):3517–3535. https://doi.org/10.1080/00207540601142645

Müller E, Engelmann J, Löffler T, Strauch J (2009) Energieeffiziente Fabriken planen und betreiben. Springer, Dordrecht, Heidelberg, London, New York

Nakajima M (2009) Evolution of material flow cost accounting (MFCA): characteristics on development of MFCA companies and significance of relevance of MFCA. Kansai Univ Rev Bus Commer 11:27–46

Negahban A, Smith JS (2014) Simulation for manufacturing system design and operation: literature review and analysis. J Manuf Syst 33(2):241–261. https://doi.org/10.1016/j.jmsy.2013.12.007

Niggeschmidt S, Moneer H, Diaz N, et al (2013) Integrating Green and sustainability aspects into life cycle performance evaluation. Green Manufacturing Sustainable Manufufacturing Partnership, UC Berkeley

Noske H, Kalogerakis C (2009) Design-to-Life-Cycle-Cost bei Investitionsgütern am Beispiel von Werkzeugmaschinen. In: Schweiger S (ed) Lebenszykluskosten optimieren. Gabler, Wiesbaden, pp 135–151

Ohlendorf M (2006) Simulationsgestützte Planung und Bewertung von Demontagesystemen. Vulkan, Essen

Ohno T (1988) Toyota production system: beyond large-scale production. CRC Press, Boca Raton

Organisation for Economic Cooperation and Development (2001) Measuring productivity – OECD manual. OECD, Paris

Patterson MG (1996) What is energy efficiency? concept, indicators and methodological issues. Energy Policy 24(5):377–390

Payne J, Bettman JR, Johnson EJ (1992) Behavioral decision research: a constructive processing perspective. Annu Rev Psychol 43(1):87–131. https://doi.org/10.1146/annurev.psych.43.1.87

Porter ME (1985) Competitive advantage: creating and sustaining superior performance. The Free Press, New York

Posselt G (2016) Towards energy transparent factories. Springer, Cham

Posselt G, Fischer J, Heinemann T et al (2014) Extending energy value stream models by the TBS dimension – Applied on a multi product process chain in the railway industry. Procedia CIRP 15:80–85. https://doi.org/10.1016/j.procir.2014.06.067

Putz M, Schlegel A, Lorenz S, et al (2011) Gekoppelte Simulation von Material- und Energieflüssen in der Automobilfertigung. Nachhalt Fabrikplan und Fabrikbetr – TBI'11 - 14 Tage des Betriebs- und Syst 135–144

Rebhan E (2002) Energiehandbuch. Springer, Berlin, Heidelberg

Reich-Weiser C, Vijayaraghavan A, Dornfeld D (2010) Appropriate use of green manufacturing frameworks. Green Manufacturing Sustainable Manufufacturing Partnership, UC Berkeley

Reinhardt S (2013) Bewertung der Ressourceneffizienz in der Fertigung. Herbert Utz Verlag, München

Reinhart G, Karl F, Krebs P (2010) Energiewertstrom: Eine Methode zur ganzheitlichen Erhöhung der Energieproduktivität. ZWF Zeitschrift für wirtschaftlichen Fabrikbetr 105(10):870–875. https://doi.org/10.1007/978-3-322-84476-7

Rommelfanger HJ, Eickemeier SH (2002) Entscheidungstheorie. Springer, Berlin, Heidelberg

Rother M, Shook J (1999) Learning to see: value stream mapping to create value and eliminate muda. Lean Enterprise Institute, Brookline

Sarkis J (1995) Manufacturing strategy and environmental consciousness. Technovation 15(2):79–97. https://doi.org/10.1016/0166-4972(95)96612-W

Sarkis J, Rasheed A (1995) Greening the manufacturing function. Bus Horiz 38(5):17–27. https://doi.org/10.1016/0007-6813(95)90032-2

Schenk M, Wirth S, Müller E (2014) Fabrikbetrieb und Fabrikplanung. Springer Vieweg, Berlin, Heidelberg

Schild U (2005) Lebenszyklusrechnung und lebenszyklusbezogenes Zielkostenmanagement: Stellung im internen Rechnungswesen, Rechnungsausgestaltung und modellgestützte Optimierung der intertemporalen Kostenstruktur. Deutscher Universitätsverlag, Wiesbaden

Schlosser R, Klocke F, Lung D (2011) Sustainabilty in manufacturing – Energy consumption of cutting processes. In: Advances in sustainable manufacturing: proceedings of the 8th global conference on sustainable manufacturing. Springer, Berlin Heidelberg, Berlin, Heidelberg, pp 85–89

Schmidt M (2007) Rohstoffe und Ressourceneffizienz - bereit für den Wettbewerb um Nachhaltigkeit. In: Statusseminar 2007. Berlin

Schönemann M (2017) Multiscale Simulation Approach for battery production systems. Springer, Cham

Schönemann M, Kurle D, Herrmann C, Thiede S (2016) Multi-product EVSM simulation. Procedia CIRP 41:334–339. https://doi.org/10.1016/j.procir.2015.10.012

Schudeleit T, Züst S, Weiss L, Wegener K (2016) The Total Energy Efficiency Index for machine tools. Energy 102:682–693. https://doi.org/10.1016/j.energy.2016.02.126

Schwartz T (1972) Rationality and the myth of the maximum. Noûs 97–117

Shim JP, Warkentin M, Courtney JF et al (2002) Past, present, and future of decision support technology. Decis Support Syst 33(2):111–126. https://doi.org/10.1016/S0167-9236(01)00139-7

Simon H (1977) The new science of management decision. Prentice Hall, Upper Saddle River

Social Hotspot Database (2019) Social Hotspot Database. https://www.socialhotspot.org/. Accessed 11 Aug 2019

Sparks D, Badurdeen F (2014) Combining sustainable value stream mapping and simulation to assess supply chain performance. In: Guan Y, Liao H (eds) Proceedings of the 2014 industrial and systems engineering research conference

Spengler T, Geldermann J, Hähre S et al (1998) Development of a multiple criteria based decision support system for environmental assessment of recycling measures in the iron and steel making industry. J Clean Prod 6(1):37–52. https://doi.org/10.1016/S0959-6526(97)00048-6

Sproedt A, Plehn J (2012) Environmental value stream map as a communicative model for discrete-event material flow simulation. In: 4th World P&OM international annual EurOMA conference 2012

Sproedt A, Plehn J, Schönsleben P, Herrmann C (2015) A simulation-based decision support for eco-efficiency improvements in production systems. J Clean Prod 105:389–405. https://doi.org/10.1016/j.jclepro.2014.12.082

Staehle W (1969) Kennzahlen und Kennzahlensysteme als Mittel der Organisation und Führung von Unternehmen. Gabler, Wiesbaden

Stahl B, Taisch M, Cannata A et al (2013) Combined Energy, material and building simulation for green factory planning. Re-engineering manufacturing for sustainability. Springer, Singapore, pp 493–498

Stiel F, Michel T, Teuteberg F (2016) Enhancing manufacturing and transportation decision support systems with LCA add-ins. J Clean Prod 110:85–98. https://doi.org/10.1016/j.jclepro.2015.07.140

Stoldt J, Putz M (2017) Procedure model for efficient simulation studies which consider the flows of materials and energy simultaneously. Procedia CIRP 61:122–127. https://doi.org/10.1016/j.procir.2016.11.195

Suh S (2005) Theory of materials and energy flow analysis in ecology and economics. Ecol Modell 189(3–4):251–269. https://doi.org/10.1016/j.ecolmodel.2005.03.011

Thiede S (2012) Energy efficiency in manufacturing systems. Springer, Berlin, Heidelberg

Thiede S, Li W, Kara S, Herrmann C (2016) Integrated analysis of energy, material and time flows in manufacturing systems. Procedia CIRP 48:200–205. https://doi.org/10.1016/j.procir.2016.03.248

Thiede S, Seow Y, Andersson J, Johansson B (2013) Environmental aspects in manufacturing system modelling and simulation – State of the art and research perspectives. CIRP J Manuf Sci Technol 6(1):78–87. https://doi.org/10.1016/j.cirpj.2012.10.004

Thies C, Kieckhäfer K, Spengler TS, Sodhi MS (2019) Operations research for sustainability assessment of products: a review. Eur J Oper Res 274(1):1–21. https://doi.org/10.1016/j.ejor.2018. 04.039

Thinkstep AG (2018) GaBi Software. https://www.gabi-software.com. Accessed 11 Aug 2019

Torres AS Jr, Gati AM (2009) Environmental value stream mapping (EVSM) as sustainability management tool. In: PICMET 2009 proceedings, 2–6 August Portland, Oregon, USA, pp 1689–1698

Turban E, Aronson JE, Liang T-P (2007) Decision support systems and intelligent systems. Prentice Hall, Upper Saddle River

United Nations Environment Programme (2011) Global guidance principles for life cycle assessment databases. Nairobi

United Nations Environment Programme (2009) Guidelines for social life cycle assessment of products. Nairobi

Verband Deuscher Maschinen-Anlagenbau e.V. (2007) VDMA 34160 - Prognosemodell für die Lebenszykluskosten von Maschinen und Anlagen. VDMA, Frankfurt am Main

Verein Deutscher Ingenieure (2005) VDI 2884 – Purchase, operating and maintenance of production equipment using Life Cycle Costing (LCC). VDI, Düsseldorf

Verein Deutscher Ingenieure (2018) VDI 3633 – Simulation of systems in materials handling, logistics and production – Terms and definitions. VDI, Düsseldorf

Verl A, Westkämper E, Abele E et al (2011) Architecture for multilevel monitoring and control of energy consumption. In: Hesselbach J, Herrmann C (eds) Glocalized solutions for sustainability in manufacturing. Springer Berlin Heidelberg, Berlin, Heidelberg, pp 347–352

Viere T, Prox M, Möller A, Schmidt M (2011) Implications of material flow cost accounting for life cycle engineering. Glocalized solutions for sustainability in manufacturing. Springer Berlin Heidelberg, Berlin, Heidelberg, pp 652–656

Walther G (2010) Nachhaltige Wertschöpfungsnetzwerke. Gabler, Wiesbaden

Weinert N, Fischer J, Posselt G, Herrrmann C (2013) Lean and green framework for energy efficiency improvements in manufacturing. In: Proceedings of the 11th global conference on sustainable manufacturing. Innovative Solutions. pp 461–466

Westkämper E (2006) Einführung in die Organisation der Produktion. Springer, Berlin, Heidelberg

Westkämper E, Zahn E (2008) Wandlungsfähige Produktionsunternehmen: Das Stuttgarter Unternehmensmodell. Springer, Berlin, Heidelberg

Widok A, Schiemann L, Jahr P, Wohlgemuth V (2012) Achieving sustainability through a combination of LCA and DES integrated in a simulation software for production processes. In: Proceedings of the winter simulation conference, pp 1219–1229

Wiendahl H-P, Nofen D, Klußmann JH, Breitenbach F (2005) Planung modularer Fabriken. Hanser, München, Wien

Wiener N (1948) Cybernetics. Sci Am 179:14–18. https://doi.org/10.2307/24945913

Wolfe P (2005) A proposed energy hierarchy. https://wolfeware.com/library/publications/Energy Hierarchy.pdf. Accessed 11 Aug 2019

Womack JP, Jones DT (1996) Lean thinking - banish waste and create wealth in your corporation. The Free Press, New York

Zein A (2013) Transition towards energy efficient machine tools. Springer, Berlin, Heidelberg

Zhou M, Pan Y, Chen Z, Yang W (2013) Optimizing green production strategies: an integrated approach. Comput Ind Eng 65(3):517–528. https://doi.org/10.1016/j.cie.2013.02.020

Zimmermann H-J, Gutsche L (1991) Multi-Criteria Analyse. Springer, Berlin, Heidelberg

Chapter 3
Existing Holistic Approaches to Increase Resource Efficiency in Manufacturing

This chapter presents an overview about existing research approaches to systematically improve resource efficiency in manufacturing. Due to the high relevance of this research topic for both scientific community and industry, various approaches have already been developed. They apply different methods and put different focuses. Because of the diversity of approaches, a rigorous definition of selection criteria is needed in order to identify relevant work in line with the research questions described in Sect. 1.2. Starting with Sect. 3.1, suitable selection criteria are derived and described. Approaches, which do not match with all criteria but are regarded as valuable contribution towards the research objectives, are briefly named. In Sect. 3.2, evaluation criteria to assess the suitability of relevant approaches for solving the challenges stated in Chap. 1 are introduced. In Sect. 3.3, the approaches are classified, described in detail and compared with respect to the fulfillment of evaluation criteria. As result, the research demand to be covered with a new approach is described in Sect. 3.4.

3.1 Selection of Relevant Approaches

A first identification of relevant approaches is conducted by applying a few general selection criteria. According to the research questions and background presented in Chaps. 1 and 2, *holistic approaches for increasing resource efficiency of transformation processes in discrete manufacturing* are selected. Accordingly, approaches from certain research domains are excluded:

- The design of products to be manufactured shall not be in the center of improvements. Consequently, sole *ecodesign and product life cycle design approaches*

© The Author(s), under exclusive license
to Springer Nature Switzerland AG 2020
S. A. Blume, *Resource Efficiency in Manufacturing Value Chains*,
Sustainable Production, Life Cycle Engineering and Management,
https://doi.org/10.1007/978-3-030-51894-3_3

like for instance Herrmann (2003), Mansour (2006) and Seow et al. (2016) are not considered. However, some of the approaches to be selected might still be able to basically assess the economic or environmental impacts of product design changes over a product's life cycle.

- Selected approaches shall be open to assess a broad range of discrete alternative solutions instead of calculating optimal values from continuous solution spaces (compare MCDA methods in Sect. 2.3.3). Hence, *operations research approaches* aiming at a mathematical optimization of clearly defined and specific problems such as route planning, inventory management, production scheduling or other analytically solvable problems are not considered. An overview of such approaches is provided by Thies et al. (2019).
- The focus of selected approaches shall be put on factory operation rather than factory planning. Accordingly, approaches that are focusing on *factory planning* (e.g. Mueller et al. 2013; Hopf 2015; Müller 2015) are not taken into account.
- The approaches shall mainly aim at value adding manufacturing activities. As a consequence, typical *supply chain management* approaches focusing on other business activities than transformation processes like supplier selection (Ude 2010; Brondi et al. 2013) or logistics management are also excluded (van der Vorst et al. 2009). Further examples from the field of sustainable supply chain management can also be found in Seuring and Müller (2008) and Seuring (2013).

As stated in Sect. 2.1.1, energy, material and media flows need to be considered for synergetic economic and environmental analyses of manufacturing systems. As a consequence, only approaches which *build on detailed quantitative material and energy flow models* of manufacturing processes are taken into account. Approaches mainly building on cost models (e.g. Zhou et al. 2013), structural factory models (e.g. Bergmann 2010), black box models (e.g. Dehning 2017) as well as concepts without clear description of underlying models (e.g. Winkler 2011) are excluded.

As described in Sect. 2.1.2, manufacturing systems are constituted by different system levels ranging from process to value chain level. For a holistic analysis, a focus on a single system level is not sufficient due to potential interactions between elements and decisions on different levels. Thus, a *consideration of multiple manufacturing system levels* is used as criterion to select suitable approaches. Approaches that are limited to process level such as Dietmair and Verl (2009), Helu et al. (2011), Kellens (2013), Niggeschmidt et al. (2013), Zein (2013) and Abele et al. (2015) are not presented in the state of research. Due to their high practical relevance for model building, they are still taken into account for the development of a new concept in Chap. 4.

Further, a *consideration of multiple evaluation dimensions* is required to face the challenges stated in Chap. 1 in an appropriate way. In practice, target conflicts between economic, environmental and technical dimensions may occur and be highly relevant for reasonable decision making. If this complexity is not reflected by approaches (e.g. Dietmair and Verl 2009; Seow and Rahimifard 2011; Lanza et al. 2013; Bleicher et al. 2014; Hopf 2015), those are not further considered in the state of research.

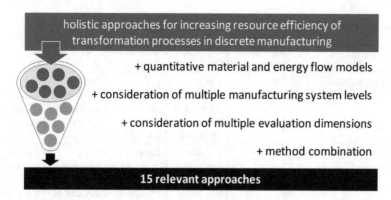

Fig. 3.1 Selection criteria to identify relevant existing approaches

Finally, a *combination of different methods for modeling and evaluation* of manufacturing activities is used as selection criterion. Hence, only innovative and methodologically broad approaches are considered. Approaches that focus on only one method such as simulation (e.g. Junge 2007; Thiede 2012; Diaz-Elsayed et al. 2013; Haag 2013; Stahl et al. 2013; Stoldt and Putz 2017) are not taken into account here.

Figure 3.1 summarizes the described selection criteria, which are applied in order to identify the most relevant approaches.

3.2 Evaluation Criteria

All selected approaches are evaluated regarding their fulfillment of a set of evaluation criteria. These criteria intend to characterize the approaches in terms of the chosen modeling focus, their ability to give decision support and their applicability in (industrial) practice. The superordinate evaluation categories, specific criteria and their fulfillment levels are introduced and described in the following. Certainly, the evaluation can only consider publicly available information.

3.2.1 Modeling

The evaluation category *modeling* is related to the overall focus of an approach. All selected approaches build upon quantitative models of energy and material flows. A characterization is conducted regarding focused *planning horizons*, *life cycle stages*, *manufacturing system levels* and *resource flows* considered in these models (compare Table 3.1).

The *planning horizon* refers to the timespans of manufacturing related business decisions, i.e. operational, tactical and strategic planning (compare Sect. 2.3.1). In

Table 3.1 Criteria and attributes of evaluation category *modeling*

Evaluation criteria		Attributes				
		○	◑	◐	◕	●
Planning horizons		–	1 of real-time, operational, tactical, strategical	2 of real-time, operational, tactical, strategical	3 of real-time, operational, tactical, strategical	4 of real-time, operational, tactical, strategical
Life cycle stages		–	1 of raw materials, production, use, end of life	2 of raw materials, production, use, end of life	3 of raw materials, production, use, end of life	4 of raw materials, production, use, end of life
Manufacturing system levels		–	1 of process, process chain, factory, value chain	2 of process, process chain, factory, value chain	3 of process, process chain, factory, value chain	4 of process, process chain, factory, value chain
Resource flows	Materials	Not considered	–	Partly considered	–	In focus
	Energies	Not considered	–	Partly considered	–	In focus
	Labor	Not considered	–	Partly considered	–	In focus

addition, real-time execution is taken into account as further time horizon, addressing real-time decision making (e.g. automatic and permanent control of machine parameters through MES systems). In an ideal case, an approach covers all time horizon horizons in order to deal with a wide range of possible manufacturing decisions.

As described in Sect. 2.4.5, a product's life cycle typically comprises the stages raw material extraction, production, use and end of life. If a manufacturing decision shall be analyzed in terms of its economic or environmental impacts, a solely evaluation of the production stage is not adequate. Due to possible problem shifting between distinct *life cycle stages* (e.g. choice of materials used in manufacturing influences costs and impacts in the other stages), a holistic consideration is advised as soon as interrelations cannot be ruled out. Thus, the coverage of multiple life cycle stages is positively appreciated in the evaluation.

Manufacturing systems can be structured into different levels. The relevance of considering multiple system levels for meaningful decision making has already been emphasized in Sect. 2.1. The consideration of multiple *manufacturing system levels* is also used as criterion to identify relevant approaches in Sect. 3.1. In addition, it is also taken into account here as evaluation criterion. The more levels an approach takes into account, the better it is assumed to allow for a holistic assessment and therefore decision making. The relevance of system levels is not weighted, as effects of a decision can entail on all system levels.

Transformation processes in manufacturing systems typically require the utilization of different *resource flows* (compare Sect. 2.1.1). Depending on the evaluation dimension, more or less of these resources might be of relevance. From an environmental point of view, only physical resource flows (energy, material) need to be considered, as they pose direct or indirect impacts on the environment. In contrast, also other resources such as human labor might be relevant for an economic evaluation. Consequently, an approach receives a better evaluation the more resources it considers. However, in practice it is usually neither reasonable nor possible to consider all resource flows, which are related to manufacturing activities. Hence, a consideration of the most relevant resources and a substantiated exclusion of minor resource flows is regarded to be sufficient.

3.2.2 Decision Support

The evaluation category *decision support* refers to the degree of support provided for the user in decision situations. An overview about derived criteria is provided in Table 3.2. As discussed in Sect. 2.3.4, systematic decision making is highly relevant for manufacturing companies and decision makers should be systematically supported, e.g. by providing concrete recommendations for action. Due to a lack of generally agreed definitions for decision support, requirements in this category have been derived from the general decision making process as described by Simon

Table 3.2 Criteria and attributes of evaluation category *decision support*

Evaluation criteria		Attributes				
		○	◔	◑	◕	●
Evaluation dimensions	Technical	Not considered	–	Partly considered	–	In focus
	Economic	Not considered	–	Partly considered	–	In focus
	Environmental	Not considered	–	Partly considered	–	In focus
	Social	Not considered	–	Partly considered	–	In focus
Reference values		No reference values	–	Internal objects	–	External objects
Identification of improvement measures		Manual identification	–	Guided	–	Automatic
Ranking of improvement measures		No ranking	–	Individual, few criteria	–	Method-based, many criteria

(1977) in Sect. 2.3.1. In an ideal case, systematic decision support is provided in all four phases, i.e. for intelligence, design, choice and implementation phase. As the actual implementation of improvement measures is regarded to be out of scope for a methodology to achieve a resource efficient manufacturing, only the first three phases are considered here.

Starting with the intelligence phase, a decision problem needs to be identified and described in a first step (Simon 1977). In terms of manufacturing, this can be realized by evaluating the manufacturing system performance using appropriate models and methods (compare Sect. 2.4). According to management theory, rather quantitative than qualitative indicators are needed to manage a system (compare Sect. 2.1.3). Thus, quantitative performance indicators should be used to assess the performance of manufacturing systems. They can be assigned to different *evaluation dimensions*, whereby technical, economic, environmental and social dimension shall be distinguished here. A consideration of multiple dimensions is regarded as basic requirement to achieve a holistic system analysis and improvement.

Furthermore, *reference values* are needed in order to judge about a manufacturing system's performance. They relate to best practice processes and systems to be used as performance benchmark (compare production functions in Sect. 2.2.2). Meaningful reference values can either be taken from internal or external sources (Deutsches Institut für Normung 2012). Internal sources can for instance refer to historic or current data of the same or similar systems, while external sources can refer to comparable competitor's systems, best available technologies or physical models. Such internal or external benchmarking can also help to identify and define the decision problem.

The target of most analyses is not only in the description of a status quo, but in the identification and evaluation of potential improvements, e.g. in order to sustain against competitors. The design phase according to Simon's decision theory describes the process of finding suitable alternative courses of action for identified problems (Simon 1977). In manufacturing, this can refer to the *identification of specific improvement measures* in terms of resource efficiency, which usually requires expert knowledge (Blume et al. 2017). In the most common case, users carry out this identification manually. Accordingly, success highly relies on a user's individual knowledge and experience. An improvement can be seen in an identification procedure which is supported by the provision of guidelines or general improvement strategies (compare strategies for sustainable manufacturing in Sect. 2.2.1). In an ideal case, improvement measures are even identified automatically and individually for a specific case.

In the choice phase of decision making processes, a solution needs to be selected. It may either consist of one single alternative or a combination of several alternatives. As selected approaches build on material and energy flow models, these solutions can typically be virtually tested in order to anticipate their impacts on the real-world system. In a simple case, consequences for the application of different alternatives are displayed beneath each other and only one or very few criteria are considered. For rather complex decision situation, individual interpretations and priority settings may then lead to contradictory conclusions and suboptimal decisions. Accordingly, decision makers should be supported through methods for a systematic *ranking of improvement measures* under consideration of many criteria. This can for instance be realized by applying multi-criteria decision analyses (MCDA) methods as introduced in Sect. 2.3.3.

3.2.3 Application

The evaluation category *application* addresses the suitability of approaches for industrial utilization. Primarily it takes into account the barriers for implementation related to limited human and financial resources, having in mind that a positive cost-benefit ratio needs to be achieved through application on the long term. Further, it assesses whether an approach is yet mature enough for industrial application or requires further developments. All applied evaluation criteria and possible attributes are subsumed in Table 3.3.

Approaches can only be systematically applied, if a clear *application procedure* is provided. It should explain the necessary steps to be performed from a user perspective, including inputs, activities, decisions as well as outcomes of each step. Depending on the type of procedure (linear or cyclical application) as well as its extent, application procedures are either classified as basic or detailed, presuming that a detailed description is more helpful in industrial application.

In order to allow for a correct operationalization of the application procedures in daily business, software implementations are almost indispensable. Thus, the *maturity of an available software implementation* is evaluated. Implementations range from solely conceptual approaches without demonstrator up to solutions, which are fully implemented in industrial environments.

The approaches presented are intended to support decision making in terms of resource efficiency in manufacturing. Hence, they can only be regarded as successful, if in total a reduction of resource utilization is achieved. Resources required to implement the approach must be taken into account in such pay off calculations. In this context, the *accessibility* of an approach plays an important role. It refers to the

Table 3.3 Criteria and attributes of evaluation category *application*

Evaluation criteria	Attributes				
	O	◔	◑	◕	●
Application procedure	Not available	–	Basic	–	Detailed
Maturity of software implementation	Software not demonstrated	Proof of concept	Tested in lab environment	Validated with industrial use cases	Fully implemented in industrial environment
Accessibility	Only concept, no implementation	Experts software and knowledge needed	Expert knowledge needed	Expert software needed	Open for non experts
Effort for implementation	High	–	Moderate	–	Low

degree of expertise needed for implementation in terms of human experts as well as specific software. Especially for SME, a lack of such know-how is often a high barrier to achieve a higher degree of resource efficiency (Herrmann et al. 2013). In the best case, an implementation requires neither highly specific expert knowledge nor expert software.

In addition, the *effort for implementation* is evaluated to consider the quantity of resources required. On the one hand, the initial effort to implement an approach is estimated, which is often related to extensive modeling and data collection activities or hardware installations. On the other hand, the continuous effort for permanent application is estimated. It is rather related to the conduction of analyses, the evaluation of improvement measures as well as the operation and maintenance of required hardware and software.

3.3 Classification, Description and Evaluation

3.3.1 Classification of Approaches

According to the selection criteria described in Sect. 3.1, 15 existing approaches are identified as relevant. All of them are model driven approaches, which evaluate resource flows in manufacturing regarding multiple evaluation dimensions. In addition, they cover more than one system level. An overview about the focuses set by the different approaches is provided in Fig. 3.2.

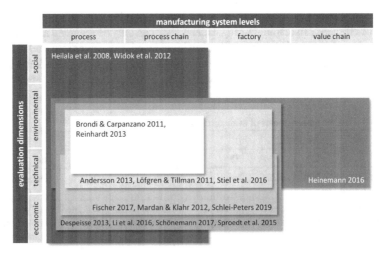

Fig. 3.2 Classification of relevant approaches in terms of covered manufacturing system levels and evaluation dimensions

3.3.2 Description of Approaches

In the following, selected approaches are described, starting with a rather narrow scope regarding system levels and evaluation dimensions. In the descriptions, the fulfillment of evaluation criteria as introduced in Sect. 3.2 is briefly discussed.

Brondi and Carpanzano presented a modular framework for LCA-based simulation of production systems in order to better exploit the potentials of LCA for manufacturing processes (Brondi and Carpanzano 2011). It builds on the description of energy and material consumption patterns for selected manufacturing processes. Accordingly, the approach follows a gate to gate modeling, considering only the production stage. Manufacturing process modules can be combined to form process chains in a DES based simulation tool, which has been supplemented with LCA features for environmental impact calculations. Technical PIs such as utilization and lead times can be received as well, while economic or social PIs are not considered. The tool can be used to assess and compare different machine configurations and use scenarios, i.e. covering tactical and strategic planning. However, decision support in terms of system improvement is not provided and the approach lacks a corresponding application procedure. A software prototype has been realized in C++ programming language. The approach was exemplified by a fictive case of a woodworking production line.

In her thesis, *Reinhardt* developed a generic approach for assessing resource efficiency in manufacturing (Reinhardt 2013). It shall provide a methodology to comparably perform LCA-like studies for manufacturing process chains. It is intended to be used in planning phase of process chains, but can also be applied to improve an existing system configuration. By modeling process chains in terms of energy and material flows, Reinhardt derives a set of indicators to express resource efficiency. Apart from PI calculations, decision support according to the defined criteria is barely existent. The concept has been implemented into a *Java* based tool named *REvalue*. A high priority has been put on usability for industrial actors, thus the tool features a strictly guided workflow and clear user interface. Further, a high degree of automation has been achieved in the tool to reduce utilization efforts. However, access to LCA databases is needed and must be established separately from the tool in order to identify impact factors for environmental assessments (compare Sect. 2.4.5). The concept application is comprehensibly described and contributes to an overall good accessibility. The feasibility of the overall approach has been proven with two rather fictive cases.

The approach presented by *Heilala et al.* evolved from the project *SIMTER* (Heilala et al. 2008). It targets at the development of an interactive tool for decision support in terms of sustainable manufacturing with particular focus on application in SME. Related work can also be found in several other publications dealing with the project (Lind et al. 2008, 2009; Johansson et al. 2009a, b, c). The approach addresses manufacturing system planning and operating issues on process and process chain levels. Different alternatives regarding work place design, production layout or production mix can be evaluated in terms of their technical, economic, environmental as well as social performance. From a methodological point of view, a discrete event model-based material flow and virtual factory simulation realized in the commercial software environments *3DCreate* and *3DRealize* are combined with analytic calculations in *Microsoft Excel*. Social aspects are addressed by applying ergonomics simulations, based on the integration of existing digital human models (Helin et al. 2007). Interfaces between the different software environments have been established and a direct embedding of the environmental assessment into the simulation tool has been achieved. While the simulation model is used to assess the dynamic behavior of the manufacturing system as well as ergonomic aspects, the analytic calculations serve to assess environmental impacts based on LCA data from public databases. Accordingly, the evaluation takes into account energy and material flows in the production stage and related impacts arising from upstream processes. Use and end of life stage are not considered. Approach and tool were applied in different industrial use cases covering machine building, automotive and chemical industry. However, a detailed application procedure could not be identified and no specific decision support is given.

An approach presented by *Widok et al.* outlines the development and application of an environmental management information system named *MILAN* that combines material flow analysis (MFA) with discrete event simulation (DES) and life cycle assessment (LCA) (Widok et al. 2012). The main objective followed is to combine different modeling and evaluation methods in order to assess manufacturing holistically. Consequently, all three dimensions of sustainability are considered in the evaluation and a balancing of competing targets is aimed. An advantage compared to most other approaches can be seen in the integration of all aspects into only one petri-net based model. Further, a direct interface to the LCA database *ecoinvent* has been established. While the focus of the simulation aspects is clearly on the production stage, the integrated LCA takes into account the whole product life cycle from cradle to grave. Decision support according to the evaluation criteria described is hardly given, but a method-based ranking of improvement measures considering multiple evaluation criteria is taken into account for future versions. Since the first version of *MILAN* was introduced in 2006, several iterations in development and applications within the projects *EcoFactory* and *EMPORER* have further extended approach and software tool (Wohlgemuth et al. 2006, 2008; Jahr et al. 2009; Page and Wohlgemuth 2010). The software development is still ongoing, currently focusing the integration of further social aspects of sustainability like physical, psychological and organizational influences on workers as well as a more user-friendly GUI (Widok and Wohlgemuth 2016; Wohlgemuth and Widok 2016). Due to a high degree of tool maturity and average efforts for implementation, the barriers for industrial application seem to be moderate. However, a detailed application procedure could not be identified.

Andersson et al. described an approach to integrate DES with LCA in a newly developed software, which evolved from the project *EcoProIT* (Andersson et al. 2011, 2012a, b; Lindskog et al. 2011; Andersson 2013). The methodology was influenced by the results of other projects such as *Reeliv* and *SIMTER*. The idea of the project was to support production engineers, who are typically not experts in modeling and simulation, in daily business. By providing a rather simple simulation tool, the engineers should be empowered to positively influence the manufacturing system's performance. Consequently, in particular improvements aiming at an operational time horizon can be tested in advance and their impacts regarding costs or emissions can be estimated. The approach bases upon a DES model, whereby all model entities are defined regarding their dynamic behavior as well as consumption patterns in terms of energy and materials. The modeling is carried out in the *EcoProIT tool*, while the actual simulation is then automatically performed in an external, commercial DES software. In principle, all stages of the product life cycle can be considered. However, due to the intended simplification of modeling and the aspired target group, the focus is put on the production stage. As a result, technical PIs like utilization rates or physical resource flows per product as well as environmental PIs can be received. Beneath the evaluation of the initial state, possible improvement measures can be modeled and assessed. Yet, the approach does neither propose a systematical application procedure to derive such measures, nor does it provide other decision support functions. Several industrial applications have already been demonstrated, e.g. for the evaluation of fork lift parts and steel can production (Andersson et al. 2012b).

Löfgren and Tillman presented an approach to combine DES with LCA in order to allow for a dynamic assessment of environmental issues in manufacturing systems (Löfgren and Tillman 2011). Related work can also be found in Löfgren (2009) and Löfgren et al. (2011). An application of the methods is proposed sequentially, hence first DES simulations need to be conducted and subsequently environmental impacts can be allocated through LCA application. The approach focuses on the modeling of the production stage considering main energy and material flows. It aims at operational planning, i.e. factors that can potentially be influenced by machine operators such as machine setup time, ramp-up or shut-down procedures, manual quality control time or tool changing intervals. As results of the analyses, PI related to the physical flow quantities per product (e.g. electricity use per product) and their inherent environmental impacts can be received. The approach focuses on the impacts caused by exemplary parameter changes, while neither concrete measures are discussed nor other decision support is given.

Stiel et al. introduced a framework and software architecture in the context of DES and LCA integration for manufacturing system and transportation modeling (Stiel et al. 2016). Motivated by the conclusion that still no commercial software fully integrates LCA and DES, the authors developed a C# based add-in for the commercial DES software *Simio* in order to access the *ecoinvent* LCA database. It enables for an assignment of LCI activities and exchanges to objects within DES models. Accordingly, no general limitations regarding the modeling exist, i.e. all manufacturing system levels and all relevant resources used in manufacturing can be covered in principle. This allows for an assessment of technical but also environmental aspects related with manufacturing activities. The approach focuses on the interactions between DES tool, LCA database and user from a technical perspective, emphasizing requirements for data management, modeling, simulation execution and result visualization. Decision support functions related to system improvements have not been described yet. The provided framework and demonstrator are rather conceived to be a first step towards the seamless implementation of future LCA add-ins into existing DES based software tools. Consequently, the actual application in an industrial context has not been proven by now and application procedures have only been provided partly. However, the seamless integration of both methods in one software environment promises to increase the efficiency of modeling and application activities.

In his thesis *Fischer* presented an approach to prioritize *solution elements* for lean and green manufacturing (Fischer 2017). Related work from the research project EMC^2-*Factory* can also be found in Weinert et al. (2013, 2014), Posselt et al. (2014) and Fischer et al. (2015). Fischer developed a solution finding process in order to identify the most promising improvement measures for a specific manufacturing system configuration from a database (compare Fig. 3.3). Solutions mainly cover challenges on operational and tactical planning horizons. His approach builds upon the energy value stream methodology (compare Sect. 2.4.1), which is further extended in terms of energy and cost related indicators in dependence on dynamic machine states. Underlying models cover factory systems with their inherent processes and TBS systems but do not address other life cycle stages. As a good usability for industrial actors was defined as one major motivation of the work, high priority was put on presenting a clear and comprehensive overall application procedure. Beneath factory modeling and value stream calculation, it also includes a weighting of user preferences as basis for a matching and individual ranking of solution elements. Accordingly, the approach enables a simple form of multi-criteria decision making. The application is supported by a stand-alone software prototype and linked SQL

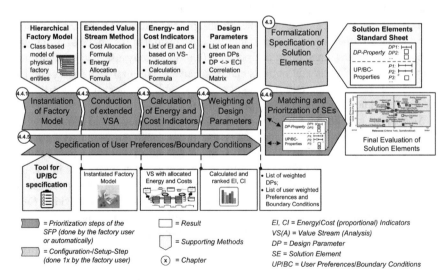

Fig. 3.3 Solution finding process for lean and green manufacturing with inherent elements (Fischer 2017)

database with a well-developed yet fragmented user interface. Modeling and application efforts are estimated to be moderate, as model elements and PIs are pre-defined in the tool and workflows appear to be intuitive and efficient. A validation of the overall approach has been conducted in a manufacturing system for train production.

Mardan and Klahr presented an approach, which combines two different software tools in order to reduce manufacturing costs in an iron foundry (Mardan and Klahr 2012). The approach builds upon the work of Thollander (2008), Solding et al. (2009), Thollander et al. (2009) and Karlsson (2011). On the one hand, simulation models created in the commercial DES software *QUEST* are used to assess the dynamics and energy flows in manufacturing systems. On the other hand, the tool *reMIND* is used to improve industrial energy systems. The authors describe an iterative procedure on how to use both tools simultaneously. It focuses on the production stage and addresses all system levels within a factory. Although resource flow consideration is limited to energy, the underlying methodology and tool have also been applied to other use cases integrating material flows and labor (e.g. Thollander et al. 2009). Technical and economic PIs are generated for system performance evaluation. Decision support is realized by providing a set of typical improvement measures. By means of the *reMIND* tool, mixed integer linear programming can be applied to mathematically optimize system parameters such as capacities or working times. However, no further decision support such as ranking of measures in terms of a multi-criteria analysis is provided. Although the application procedure is described thoroughly, the barriers for industrial application still seem to be significant. On the one hand, a full integration of both software tools has not been physically reached. A high degree of expert knowledge is, on the other hand, needed for practical implementation, while only a basic application procedure is provided.

Schlei-Peters developed a concept for decision making in the production stage regarding invests for environmental protection measures (Schlei-Peters 2019). Related work can be found in Schlei-Peters et al. (2015, 2018). The approach is designed for tactical decision making, but may also give indications for strategic decisions. It builds on the MEFA method, modeling relevant system elements and resource flows (materials and energies) from process level up to entire production systems including TBS. The system performance is basically assessed using resource flow quantities as technical indicators. Although the approach is dedicated to environmental protection measures, an environmental evaluation is argued based on resource flow quantities only and no in-depth analysis in terms of environmental impacts is conducted. Potential invests into improvement measures are evaluated regarding their economic benefits in terms of net present values and payback periods. Decision making is supported by means of a mathematical optimization model, which is used to rank measures according to these performance indicators. Further criteria such as the reduction of a specific material or energy flow can be considered as constraints in the optimization model. A concept implementation was conducted in *Microsoft Excel* (MEFA models) as well as *AIMMS* and *CPLEX* (mathematical decision models). While the concept is basically applicable for all production systems and a general application procedure is provided, modeling and implementation focus on a specific case study to assess industrial cooling systems. Effort and expert knowledge to implement the approach are regarded to be at least moderate, as different software tools are used and the transfer to other systems appears to be challenging.

The approach shown by *Despeisse* focuses on *sustainable manufacturing tactics and improvements* on different manufacturing system levels, covering processes, facilities and whole factories (Despeisse 2013). Related work can also be found in Despeisse et al. (2012a, b, 2013), Oates et al. (2012) and Ball et al. (2013). The approach is strongly related to the research project *THERM*, which aimed at the development of a modeling and simulation tool to support manufacturing plant design and improvement (Ball et al. 2011, 2012). Despeisse provides a general improvement methodology and application procedure to support companies in measure identification and implementation. The approach targets at the production stage, taking the factory gates as system boundaries. Improved building design and manufacturing processes are in focus, taking into account physical resource flows such as material, energy and water. Consequently, the approach allows for technical, environmental and economic assessments. General improvement hierarchies as well as a tactics library can be exploited for measure identification (compare Fig. 3.4). The library contains more than 1.000 specific improvement measures taken from theory and practice, covering operational up to strategic planning issues. However, further methodological decision support is not provided, hence target conflicts may be identified but cannot be solved systematically. A software demonstrator has been developed by a professional software developer and tested in at least four industrial use cases within the project. Required knowledge and efforts for implementation are still regarded to be high due to the complexity of models to be set up.

Li et al. developed a sustainability cockpit to allow companies, especially SME, to continuously assess and improve their sustainability performance (Li et al. 2016). Related work can also be found in Alvandi et al. (2015, 2016), Thiede et al. (2016a) and Alvandi et al. (2017). The approach is based on a DES factory model in the simulation software *AnyLogic* with a link to enterprise resource planning systems (compare Fig. 3.5). The degree of modeling and simulation depth is flexible, as three different data sourcing strategies were implemented, whereby missing data

	Resource	Technology
Manage	**A1 Manage Resource** A11 Align resource input profile with production schedule A12 Optimise production schedule to improve efficiency A13 Optimise resource input profile to improve efficiency A14 Synchronise waste generation and resource demand to allow reuse A15 Waste collection, sorting, recovery and treatment	**A3 Manage Technology** A31 Repair and maintain A32 Change set points/running load, reduce demand A33 Switch off/standby mode when not in use A34 Monitor performance A35 Control performance
Change	**A2 Change Resource** A21 Remove unnecessary resource usage A22 Replace resource input for better one A23 Add high efficiency resource A24 Reuse waste output as resource input A25 Change resource flow layout	**A4 Change Technology** A41 Remove unnecessary technology A42 Replace technology for better one A43 Add high efficiency technology A44 Change the way the function is accomplished A45 Change technology layout

Fig. 3.4 Excerpt of tactics library for sustainable manufacturing (Despeisse 2013)

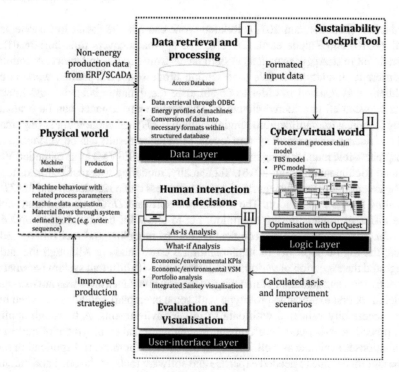

Fig. 3.5 System concept for a sustainability cockpit (Li et al. 2016)

can be filled using pre-defined process models. The user interface is realized in *Microsoft Excel*, covering data input, scenario configuration as well as result visualization, including an economic and environmental VSM (E2VSM) representation of the process chain (Alvandi et al. 2016). The simulation model can be exported as *Java* applet and therefore serve as stand-alone model. In general, the approach aims at giving strategic support for the plant management in terms of sustainability related decisions. Regarding the scope, a factory with all its relevant subsystems such as processes, TBS and building shell is considered. Sustainability is assessed in terms of economic and environmental dimensions, taking into account relevant material and energy flows. Decision support according to the criteria defined is given by the implementation of a multi-objective optimization module, using weighted sums to rank competing improvement scenarios. The actual optimization is carried out using the commercial software *OptQuest*. However, no support is provided to identify improvement measures or to benchmark the current system performance. Due to a possibility to bridge data gaps, a high degree of usability and a seamless integration with ERP systems, barriers for continuous industrial application appear to be moderate after initial implementation. The concepts applicability has already been proven in at least three industrial use cases.

Schönemann presented a multiscale simulation approach for a holistic evaluation and improvement of multi-level manufacturing systems with a focus on battery production (Schönemann 2017). Related work can also be found in Schönemann et al. (2015) and Thiede et al. (2016b). The approach covers time horizons from operational to strategic as well as system levels from process to factory. A coupling of different simulation models representing processes, TBS, human workers and buildings was realized in order to co-simulate the dynamic behavior and interactions between all considered elements. The system performance can be evaluated using technical (e.g. utilization), environmental (CO_2) and economic PIs (e.g. material and energy costs) on all system levels. Simulation results can be visualized in a comprehensible manner using multi-product EVSM (MESVM) as an extended VSM method (Schönemann et al. 2016). Technically, modeling was realized in different software environments such as *AnyLogic* for process chain simulation and *MATLAB Simulink* for TBS simulation. The middleware software *TISC* was used to synchronize different models during execution, i.e. to allow for communication and data transfer. A focus was put on technical aspects of modeling and realization, underlining the high complexity of implementing a co-simulation. Although the author suggested the execution of various simulation runs with different system parameters, the derivation and selection of meaningful improvement measures was not examined in detail. A first extendable prototype with some exemplary modules has been built and successfully validated with data from lab environments. A thorough application procedure to support development, maintenance and employment of multiscale simulations is available as well. However, the barriers for industrial application seem to be significant due to needed expertise and software tools on the one hand and high efforts for initial model set up and data collection on the other hand.

In order to holistically evaluate the eco-efficiency of manufacturing with reasonable efforts, *Sproedt et al.* presented an approach of a simulation based decision support to improve production systems (Sproedt et al. 2013, 2015). It aims at a seamless integration of LCA with DES, embedded into a structured procedure to derive and apply improvement measures (compare Fig. 3.6). The authors designed and implemented a new prototypical simulation software with direct access to the *ecoinvent* LCA database. It builds on the open-source plugin framework *EMPINIA* introduced by Jahr et al. (2009). A stepwise procedure is proposed to aid decision makers during system analysis and modeling activities. Environmental VSM according to Plehn et al. (2012) and Sproedt and Plehn (2012) is used as a basis to structure and describe the transformation processes along process chains. Before modeling a manufacturing system, all processes need to be rated by experts regarding eventual impacts and expected improvement potentials. The focus during modeling and data acquisition can then be set accordingly. Models consider all (relevant) material and energy flows as well as human labor. As results, various PIs are displayed within an evaluation module covering technical, economic and environmental dimensions. The authors give decision support for measure identification by providing an overview about general focus areas (e.g. reduce impact of energy or materials) with suitable improvement strategies (e.g. substitute material). However, specific measures for direct application are not given and a ranking of tested measures is not discussed. Due to the broadness of the approach, consequences resulting from manifold improvement scenarios can be assessed. The demonstrated cases mainly

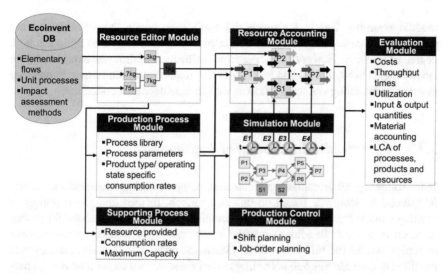

Fig. 3.6 Simulation approach for eco-efficiency assessments by integrating DES and LCA (Sproedt et al. 2013, 2015)

refer to rather operational tasks on shop floor level like production planning and control (e.g. shift planning, product sequence planning). Additionally, design and use of technical equipment are addressed, e.g. by elimination of bottlenecks and over capacities. Further, the approach is also feasible to address tactical and strategic decisions including product design. To increase the usability, modeling support is provided by the tool and all functionalities are integrated into one software environment. Applicability and user-friendliness have been ensured by testing the modeling procedure with non-experts and through applications in at least four industrial use cases.

Heinemann presented an approach to increase energy and resource efficiency in the aluminum die casting industry, which evolved from the research project *ProGRess* (Heinemann 2016). Related work can be found in Verl et al. (2011) and Heinemann et al. (2012, 2013). Motivated by the significant environmental impacts presumed for aluminum production, a methodological support was developed to allow for holistic improvements along metal casting value chains. The scope considered is very broad, reaching from process to value chain level. Different modeling and simulation methods comprising MEFA, LCA and simulation have been combined in order to cover the evaluation of improvement scenarios on different manufacturing system levels. Except from the product use stage, the entire product life cycle is considered. The approach intends to cover all relevant physical flows regarding energy and materials, deriving PIs from technical and environmental perspective. Still, decision support in terms of the criteria defined is hardly given. A detailed application procedure stresses the evaluation of improvement scenarios, but without providing clear guidance on how to identify suitable alternatives. Heinemann applied his approach to different metal casting value chains but also demonstrated a use case from battery module assembly. Due to the successful application within the project and other presented use cases, the software demonstrator can be assumed to feature a high degree of maturity. However, the concurrent utilization of different expert software environments like *Umberto* and *MAGMASOFT* requires profound expert knowledge in terms of modeling and application as well as extensive data collection.

3.3.3   Comparative Evaluation of Approaches

In the following, all presented approaches are compared using the evaluation criteria introduced in Sect. 3.2. Based on this comparison, focuses and shortcomings of existing work in the addressed field of research are discussed and demand for further research is derived. To achieve a better comparability of results, a quantification is carried out for the fulfillment of all evaluation criteria. 0.25 points are assigned per filled quarter *Harvey ball* according to the presented evaluation schemes. Consequently, every approach reaches a fulfillment level between 0 and 1 for each criterion. By taking the numerical average over all criteria per approach, its overall suitability for a holistic assessment and improvement of resource efficiency in manufacturing according to the criteria defined can be expressed. In addition, the average fulfillment for each criterion is calculated in order to identify aspects that may need further investigation. A weighting of criteria is not performed. The entire evaluation is carried out to the best of the author's knowledge using publicly available references. Consequently, the overall score must neither be understood as a valid rating about the quality nor about the scientific or industrial relevance of an approach. A summary of the evaluation results is provided in Table 3.4.

From the overview, the following observations can be stated:

- No approach does completely fulfil the evaluation criteria defined. The highest average score of 0.64 is reached by the approach of Sproedt et al. (2015).
- Several evaluation criteria are in general well addressed. This applies for most criteria from the evaluation category *modeling* and partly for the evaluation category *application*. In contrast to that, shortcomings can be stated mainly for criteria from the evaluation category *decision support*.
- Regarding system *modeling*, many approaches use well-established methods for manufacturing system modeling and evaluation such as MEFA, VSM, LCA and simulation (compare Sect. 2.4). Many holistic approaches in terms of *planning horizon* and *manufacturing system levels* are identified. However, the approach of Heinemann (2016) is the only one explicitly addressing the value chain level of manufacturing. *Resource flows* in terms of materials and energies are also very well covered, whereas human labor is hardly considered. Regarding product *life cycle stages* a focus is usually put on the production stage. Some approaches do also integrate upstream processes related to the extraction of raw materials or production of semi-finished products, while the entire product life cycle is only covered by Widok et al. (2012) and Andersson (2013).

Table 3.4 Overview about evaluation results for relevant approaches

		Brondi and Carpanzano (2011)	Reinhardt (2013)	Heilala et al. (2008)	Widok et al. (2012)	Andersson (2013)	Löfgren and Tillman (2011)	Stiel et al. (2016)	Fischer (2017)	Mardan and Klahr (2012)	Schlei-Peters (2019)	Despeisse (2013)	Li et al. (2016)	Schönemann (2017)	Sproedt et al. (2015)	Heinemann (2016)	Average
Modeling																	
Planning horizons		◐	◐	◐	◐	◑	◑	◐	◐	◐	◐	◐	◐	◐	◐	◐	0.60
Life cycle stages		◑	◐	◐	●	●	●	●	◑	◐	○	◐	◑	◐	◐	●	0.45
Manufacturing system levels		◐	◐	◐	◐	◐	◐	◐	◐	◐	◐	◐	◐	◐	◐	●	0.70
Resource flows	Materials	●	●	●	●	●	●	●	●	●	●	●	●	●	●	●	0.93
	Energies	●	●	●	●	●	●	●	●	◑	●	●	●	●	●	●	0.97
	Labor	○	○	◑	○	○	○	○	○	◐	○	○	○	●	○	◐	0.17
Decision support																	
Evaluation dimensions	Technical	●	●	●	●	●	◑	●	●	●	◑	◑	●	●	●	●	0.90
	Economic	○	○	◑	●	○	○	●	●	◑	●	◑	◑	●	●	○	0.47
	Environmental	●	●	●	●	◐	◑	●	○	○	○	◑	◑	◑	●	●	0.63
	Social	○	○	◑	◑	○	○	○	○	○	○	○	○	○	○	○	0.07
Reference values		○	○	○	○	○	○	○	○	○	○	○	○	○	◑	○	0.07
Identification of improvement measures		○	○	○	○	○	○	○	○	◐	○	◐	○	○	◐	○	0.20
Ranking of improvement measures		○	○	○	○	○	○	○	◐	○	◐	◐	●	○	○	○	0.17
Application																	
Application procedure		○	●	◑	○	○	○	○	●	◑	●	●	○	●	●	●	0.53
Maturity of software demonstrator		◑	◑	◑	◑	◐	○	◑	◑	◑	◑	◑	◑	◐	◐	◑	0.52
Accessibility		◐	●	◑	◐	◐	◑	◑	◑	◑	◑	○	●	◑	◐	◑	0.45
Effort of implementation		◐	◑	◑	◐	◐	○	◑	◐	○	◑	○	◑	○	◐	○	0.33
Total																	
	Average	0.39	0.50	0.51	0.54	0.44	0.35	0.49	0.50	0.44	0.43	0.50	0.56	0.54	0.64	0.57	0.49

- In terms of *decision support*, all approaches take into account multiple *evaluation dimensions*. Technical indicators are calculated within all approaches to express the operational manufacturing system performance or quantitative energy and material flows. The consideration of economic and environmental aspects is also common, while social aspects are usually neglected except than by Heilala et al. (2008) and Widok et al. (2012). Reasons for this may be seen in the relatively new domain of social LCA (compare Sect. 2.4.5). Further, an assessment of social aspects cannot be performed on the same data basis like environmental or economic assessments, hence additional data acquisition would be required (United Nations Environment Programme 2009; Benoît-Norris et al. 2011). Consequently, social aspects considered mainly concentrate on aspects within the factory gates such as ergonomics or direct threads at the workplace. In contrast to the detailed performance evaluation provided by most approaches, decision support in terms of system improvement is barely given. Most approaches fail to adequately support users in system performance comparison, e.g. by the provision of meaningful *reference values*. As a result, significant improvement potentials may remain hidden in many cases. Only Sproedt et al. (2015) propose to use expert consultations in order to identify processes that might feature high improvement potentials, while Heinemann (2016) provides case specific data to develop a reference value chain. Further, the *identification of improvement measures* is usually left to the user. Here, only Fischer (2017) presents an advanced approach, providing a software tool to automatically identify suitable measures from a measure database. Few other approaches provide at least some general guidelines to identify promising alternatives. A systematic *ranking of improvement measures* is also hardly given. Only Li et al. (2016) apply advanced mathematical algorithms to find good measures in accordance with user preferences for multi-criteria decision making.
- In terms of application, *application procedures* are provided by some approaches in a very detailed form, while for others no instructions or only very general information is available. The provision of a detailed application procedure often goes along with the existence of a mature *software demonstrator*. This is meaningful due to the complexity of most approaches. Hence, they cannot easily be applied using available commercial software tools only. In any case, a thorough application procedure in combination with a mature software tool to follow the procedure highly fosters the suitability for continuous industrial application. As many approaches have emerged from larger projects, involving both research institutions and industrial actors, corresponding software implementations could already be tested and validated in these settings. Regarding the criteria *accessibility* and *effort of utilization* most approaches do not receive a rating higher than 0.5. This is related to required expert knowledge and expert software as well as significant efforts related with data collection. Especially for SME, a (full) implementation often seems to be unrealistic. Only the approach of Reinhardt (2013) appears to be generally applicable without extensive expert knowledge or software. In terms of a continuous application, the approach of Li et al. (2016) promises comparably low efforts due to its seamless integration with ERP systems.

Based on these observations, further conclusions regarding existing research demands are derived in the following.

3.4 Derivation of Research Demand

From the comparative overview in Sect. 3.3.3 it can be concluded that several criteria are already well fulfilled and do not necessitate fundamental research. In particular, a holistic modeling and evaluation of manufacturing systems within the scope of the factory gates for operational, tactical and strategic time horizons can be regarded as state of the art. Accordingly, a new approach can build on existing work in terms of holistic modeling and evaluation of manufacturing systems. The research demand described in the following evolves from the identification of shortcomings in existing approaches such as a general lack of decision support. Overcoming these gaps could foster the ability of (industrial) actors to assess and improve manufacturing systems in a more holistic and user-friendly way. Accordingly, a new concept shall in particular close the following gaps:

Most approaches are limited to the evaluation of the production stage. This limitation to a *gate to gate scope is susceptible for problem shifting along the product life cycle* (compare Sect. 2.4.5). Thus, robust statements about global advantages and drawbacks of decisions cannot be derived. As an example, Despeisse (2013) justifies the selection of a gate to gate scope with the limited control that some companies have on activities outside the factory gates. However, the consideration of upstream processes is regarded as indispensable, as major environmental impacts can often be traced back to raw material extraction and processing. Also, downstream processes should not be generally neglected, in particular if the effects of product design changes or end of life treatment are of interest. Future research shall therefore aim at cradle to grave instead of gate to gate modeling and evaluation.

To avoid problem shifting between departments within a factory or actors in the same value chain, different manufacturing levels shall be taken into account. In some cases, local improvements for one actor might induce drawbacks for another actor. As most existing approaches *do not sufficiently consider actor spanning activities along value chains*, they tend to foster local rather than global improvements. In this context, Heinemann was capable to demonstrate that a liquid transport of aluminum can lead to a significant global reduction of resource demands, although the local consequences for alloy suppliers are negative (Heinemann 2016). Future research shall take into account interdependencies of activities on different manufacturing system levels and in particular consider impacts on value chain level.

Major shortcomings identified in existing approaches are related to decision support aspects, hence research shall put a stronger focus on actively supporting users during the interpretation of results and derivation of proper manufacturing decisions. For most existing approaches, a support for the *interpretation of results and drawing of conclusions is regarded to be insufficient*. Decision quality is thus highly depending on individual knowledge and experiences. Three main questions arise in this context, which reflect the three phases of Simon's decision making process as presented in Sect. 2.3.2 (Simon 1977):

1. Where are the *greatest levers* to apply changes? (problem definition; intelligence phase)
2. What are *measures* to improve the system performance? (identification of alternatives; design phase)
3. Which measures promise the highest *improvement potentials*? (selection of alternatives; choice phase).

The first question refers to the system elements and resource flows which may feature improvement potentials. In most approaches identified, *users are hardly supported to find the best levers for improvements*. Consequently, they are in danger of concentrating exclusively on processes or resources with the worst performance in terms of operational behavior or resource demands (e.g. bottleneck process, highest energy demand, highest total costs), while main saving potentials might remain hidden at other spots. Future research shall enable for a better comparability of system performance indicators in order to find the most promising levers more easily.

The second question refers to a stated *lack of support during measure identification*, leading to a high reliability on expert involvement into evaluation processes. Consequently, the need for expert consultation shall be reduced in order enable quicker and more efficient decision processes.

The third question relates to the selection of best alternatives in accordance with followed targets. If multiple evaluation dimensions are considered, target conflicts can arise from the implementation of changes, e.g. cost reductions at the expense of higher environmental impacts. Although the approaches presented cover multiple evaluation dimensions, most of them *do not provide support to solve target conflicts arising from multi-criteria evaluation*. In practice, an identification of measures providing the best use value in accordance with the individual objectives of an actor cannot be ensured. Adequate methods for multi-criteria decision analysis (MCDA) are well-established in practice (compare Sect. 2.3.3). Hence, they shall be integrated into overall concepts and application procedure.

Finally, significant *barriers to implement most approaches in industry are identified*. Both commercial software and extensive expert knowledge in terms of system modeling and simulation are typically needed. New approaches shall go along with a better accessibility for industry. The approaches of Reinhardt (2013) and Li et al. (2016) prove, that accessibility can be fostered by reducing the reliance on (expensive) commercial expert software tools. Further, the efforts for both initial setup but also for continuous application shall be reduced in comparison to existing approaches. A reduced initial modeling effort can for instance be reached by providing pre-defined system element models as demonstrated by Heilala et al. (2008) and Widok et al. (2012).

References

Abele E, Braun S, Schraml P (2015) Holistic simulation environment for energy consumption prediction of machine tools. Procedia CIRP 29:251–256. https://doi.org/10.1016/j.procir.2015.02.059

Alvandi S, Bienert G, Li W, Kara S (2015) Hierarchical modelling of complex material and energy flow in manufacturing systems. Procedia CIRP 29:92–97. https://doi.org/10.1016/j.procir.2015.01.023

Alvandi S, Li W, Schönemann M et al (2016) Economic and environmental value stream map (E2VSM) simulation for multi-product manufacturing systems. Int J Sustain Eng 9(6):354–362. https://doi.org/10.1080/19397038.2016.1161095

Alvandi S, Li W, Kara S (2017) An integrated simulation optimisation decision support tool for multi-product production systems. Mod Appl Sci 11:56. https://doi.org/10.5539/mas.v11n6p56

Andersson J (2013) Life cycle assessment in production flow simulation for production engineers. In: Proceedings of the 22nd international conference on production research

Andersson J, Skoogh A, Johansson B (2011) Environmental activity based cost using discrete event simulation. In: 2011 winter simulation conference, WSC 2011, pp 891–902

Andersson J, Johansson B, Berglund J, Skoogh A (2012a) Framework for ecolabeling using discrete event simulation. In: Emerging M&S applications in industry and academia symposium 2012, EAIA 2012–2012 spring simulation multiconference

Andersson J, Skoogh A, Johansson B (2012b) Evaluation of methods used for life-cycle assessments in discrete event simulation. In: Proceedings of the 2012 winter simulation conference (WSC). IEEE, pp 1–12

Ball P, Despeisse M, Evans S et al (2011) Modelling energy flows across buildings, facilities and manufacturing operations. In: Proceedings of the 28th international manufacturing conference (IMC28), pp 290–297

Ball PD, Despeisse M, Evan S et al (2012) Modelling buildings, facilities and manufacturing operations to reduce energy consumption. In: Proceedings of the POMS 23rd annual conference, pp 26–28

Ball PD, Despeisse M, Evans S et al (2013) Factory modelling: combining energy modelling for buildings and production systems. IFIP Adv Inf Commun Technol 397:158–165. https://doi.org/10.1007/978-3-642-40352-1_21

Benoît-Norris C, Vickery-Niederman G, Valdivia S et al (2011) Introducing the UNEP/SETAC methodological sheets for subcategories of social LCA. Int J Life Cycle Assess 16(7):682–690. https://doi.org/10.1007/s11367-011-0301-y

Bergmann L (2010) Nachhaltigkeit in Ganzheitlichen Produktionssystemen. Vulkan Verlag, Essen

Bleicher F, Duer F, Leobner I et al (2014) Co-simulation environment for optimizing energy efficiency in production systems. CIRP Ann Manuf Technol 63(1):441–444. https://doi.org/10.1016/j.cirp.2014.03.122

Blume S, Kurle D, Herrmann C, Thiede S (2017) Toolbox for increasing resource efficiency in the European metal mechanic sector. Procedia CIRP 61:40–45. https://doi.org/10.1016/j.procir.2016.11.247

Brondi C, Carpanzano E (2011) A modular framework for the LCA-based simulation of production systems. CIRP J Manuf Sci Technol 4(3):305–312. https://doi.org/10.1016/j.cirpj.2011.06.006

Brondi C, Fornasiero R, Vale M et al (2013) Modular framework for reliable LCA-based indicators supporting supplier selection within complex supply chains. In: IFIP international conference on advances in production management systems. Springer, pp 200–207

Dehning P (2017) Steigerung der Energieeffizienz von Fabriken der Automobilproduktion. Springer Fachmedien, Wiesbaden

Despeisse M (2013) Sustainable manufacturing tactics and improvement methodology: a structured and systematic approach to identify improvement opportunities. Cranfield University

Despeisse M, Ball PD, Evans S, Levers A (2012a) Industrial ecology at factory level – a conceptual model. J Clean Prod 31(10):30–39. https://doi.org/10.1016/j.jclepro.2012.02.027

Despeisse M, Mbaye F, Ball PD, Levers A (2012b) The emergence of sustainable manufacturing practices. Prod Plan Control 23(5):354–376. https://doi.org/10.1080/09537287.2011.555425

Despeisse M, Oates MR, Ball PD (2013) Sustainable manufacturing tactics and cross-functional factory modelling. J Clean Prod 42:31–41. https://doi.org/10.1016/j.jclepro.2012.11.008

Diaz-Elsayed N, Jondral A, Greinacher S et al (2013) Assessment of lean and green strategies by simulation of manufacturing systems in discrete production environments. CIRP Ann 62(1):475–478. https://doi.org/10.1016/j.cirp.2013.03.066

Dietmair A, Verl A (2009) A generic energy consumption model for decision making and energy efficiency optimisation in manufacturing. Int J Sustain Eng 2(2):123–133. https://doi.org/10.1080/19397030902947041

Deutsches Institut für Normung (2012) DIN EN 16231 – Energieeffizienz-Benchmarking-Methodik

Fischer J (2017) Prioritizing components for lean and green manufacturing. Vulkan, Essen

Fischer J, Weinert N, Herrmann C (2015) Method for selecting improvement measures for discrete production environments using an extended energy value stream model. Procedia CIRP 26:133–138. https://doi.org/10.1016/j.procir.2014.07.100

Haag H (2013) Eine Methodik zur modellbasierten Planung und Bewertung der Energieeffizienz in der Produktion. Fraunhofer Verlag, Stuttgart

Heilala J, Vatanen S, Tonteri H et al (2008) Simulation-based sustainable manufacturing system design. In: 2008 winter simulation conference. IEEE, pp 1922–1930

Heinemann T (2016) Energy and resource efficiency in aluminium die casting. Springer International Publishing, Cham

Heinemann T, Machida W, Thiede S et al (2012) A hierarchical evaluation scheme for industrial process chains: aluminum die casting. In: 19th CIRP international conference on life cycle engineering, Berkeley. Springer, Berlin, Heidelberg, pp 503–508

Heinemann T, Thiede S, Herrmann C (2013) Handlungsfeld Bewertung von Energie- und Ressourceneffizienz in industriellen Prozessketten. In: Energie- und ressourceneffiziente Produktion von Aluminiumdruckguss. Springer, Berlin, Heidelberg, pp 277–320

Helin K, Viitaniemi J, Aromaa S et al (2007) OSKU – digital human model in the participatory design approach. VTT Work Pap 83:1–37

Helu M, Rühl J, Dornfeld D et al (2011) Evaluating trade-offs between sustainability, performance, and cost of green machining technologies. In: Glocalized solutions for sustainability in manufacturing. Proceedings of the 18th CIRP international conference on life cycle engineering, pp 195–200

Herrmann C (2003) Unterstützung der Entwicklung recyclinggerechter Produkte. Vulkan, Essen

Herrmann C, Posselt G, Thiede S (2013) Energie- und Hilfsstoffoptimierte Produktion. Springer Vieweg, Berlin, Heidelberg

Hopf H (2015) Methodik zur Fabriksystemmodellierung im Kontext von Energie- und Ressourceneffizienz. Springer Fachmedien, Wiesbaden

Jahr P, Schiemann L, Wohlgemuth V (2009) Development of simulation components for material flow simulation of production systems based on the plugin architecture framework EMPINIA. In: Proceedings of the EnviroInfo 2009, Berlin, pp 151–159

Johansson B, Fasth A, Stahre J et al (2009a) Enabling flexible manufacturing systems by using level of automation as design parameter. In: Proceedings of the 2009 winter simulation conference (WSC), pp 2176–2184

Johansson B, Kacker R, Kessel R et al (2009b) Utilizing combinatorial testing on discrete event simulation models for sustainable manufacturing. In: Proceedings of the 14th design for manufacturing and the life cycle conference DFMLC14. ASME, pp 1095–1101

Johansson B, Skoogh A, Mani M, Leong S (2009c) Discrete event simulation to generate requirements specification for sustainable manufacturing systems design. In: Proceedings of the 9th workshop on performance metrics for intelligent systems – PerMIS'09. ACM Press, New York, p 38

Junge M (2007) Simulationsgestützte Entwicklung und Optimierung einer energieeffizienten Produktionssteuerung. Kassel University Press, Kassel

Karlsson M (2011) The MIND method: a decision support for optimization of industrial energy systems – principles and case studies. Appl Energy 88(3):577–589. https://doi.org/10.1016/j.apenergy.2010.08.021

Kellens K (2013) Energy and resource efficient manufacturing – unit process analysis and optimisation. KU Leuven

Lanza G, Greinacher S, Jondral A, Moser R (2013) Monetary assessment of an integrated lean-/green-concept. In: Proceedings of the 11th global conference on sustainable manufacturing GCSM, pp 548–553

Li W, Alvandi S, Kara S et al (2016) Sustainability cockpit: an integrated tool for continuous assessment and improvement of sustainability in manufacturing. CIRP Ann Manuf Technol 65(1):5–8. https://doi.org/10.1016/j.cirp.2016.04.029

Lind S, Krassi B, Johansson B et al (2008) SIMTER: a production simulation tool for joint assessment of ergonomics, level of automation and environmental impacts. In: The 18th international conference on flexible automation and intelligent manufacturing, pp 1025–1031

Lind S, Johansson B, Stahre J et al (2009) SIMTER – a joint simulation tool for production development. VTT Work Pap 125:1–49

Lindskog E, Lundh L, Berglund J et al (2011) A method for determining the environmental footprint of industrial products using simulation. In: Proceedings of the 2011 winter simulation conference, pp 2136–2147

Löfgren B (2009) Capturing the life cycle environmental performance of a company's manufacturing system. Chalmers University of Technology

Löfgren B, Tillman AM (2011) Relating manufacturing system configuration to life-cycle environmental performance: discrete-event simulation supplemented with LCA. J Clean Prod 19(17–18):2015–2024. https://doi.org/10.1016/j.jclepro.2011.07.014

Löfgren B, Tillman AM, Rinde B (2011) Manufacturing actor's LCA. J Clean Prod 19(17–18):2025–2033. https://doi.org/10.1016/j.jclepro.2011.07.008

Mansour M (2006) Informations- und Wissensbereitstellung für die lebenszyklusorientierte Produktentwicklung. Vulkan, Essen

Mardan N, Klahr R (2012) Combining optimisation and simulation in an energy systems analysis of a Swedish iron foundry. Energy 44(1):410–419. https://doi.org/10.1016/j.energy.2012.06.014

Mueller F, Cannata A, Stahl B et al (2013) Green factory planning. In: IFIP international conference on advances in production management systems, pp 167–174. https://doi.org/10.1007/978-3-642-41266-0_21

Müller F (2015) Modular planning concept for green factories. Vulkan-Verlag, Essen

Niggeschmidt S, Moneer H, Diaz N et al (2013) Integrating green and sustainability aspects into life cycle performance evaluation. Green Manuf Sustain Manuf Partnersh

Oates MR, Despeisse M, Ball PD et al (2012) Design of sustainable industrial systems by integrated modelling of factory building and manufacturing processes. In: Proceedings of the 10th CIRP global conference on sustainable manufacturing – towards implementing sustainable manufacturing, pp 1–8

Page B, Wohlgemuth V (2010) Advances in environmental informatics: integration of discrete event simulation methodology with ecological material flow analysis for modelling eco-efficient systems. Procedia Environ Sci 2:696–705. https://doi.org/10.1016/j.proenv.2010.10.079

Plehn J, Sproedt A, Gontarz A, Reinhard J (2012) From strategic goals to focused eco-efficiency improvement in production – bridging the gap using environmental value stream mapping. In: 10th global conference on sustainable manufacturing (GCSM 2012)

Posselt G, Fischer J, Heinemann T et al (2014) Extending energy value stream models by the TBS dimension – applied on a multi product process chain in the railway industry. Procedia CIRP 15:80–85. https://doi.org/10.1016/j.procir.2014.06.067

Reinhardt S (2013) Bewertung der Ressourceneffizienz in der Fertigung. Herbert Utz Verlag, München

Schlei-Peters I (2019) Modellbasierte Investitionsplanung produktionsbezogener Umweltschutzmaßnahmen. Springer Fachmedien, Wiesbaden

Schlei-Peters I, Kurle D, Wichmann MG et al (2015) Assessing combined water-energy-efficiency measures in the automotive industry. Procedia CIRP 29:50–55. https://doi.org/10.1016/j.procir. 2015.02.013

Schlei-Peters I, Wichmann MG, Matthes IG et al (2018) Integrated material flow analysis and process modeling to increase energy and water efficiency of industrial cooling water systems. J Ind Ecol 22:41–54. https://doi.org/10.1111/jiec.12540

Schönemann M (2017) Multiscale simulation approach for battery production systems. Springer International Publishing, Cham

Schönemann M, Herrmann C, Greschke P, Thiede S (2015) Simulation of matrix-structured manufacturing systems. J Manuf Syst 37:104–112. https://doi.org/10.1016/j.jmsy.2015.09.002

Schönemann M, Kurle D, Herrmann C, Thiede S (2016) Multi-product EVSM simulation. Procedia CIRP 41:334–339. https://doi.org/10.1016/j.procir.2015.10.012

Seow Y, Rahimifard S (2011) A framework for modelling energy consumption within manufacturing systems. CIRP J Manuf Sci Technol 4(3):258–264. https://doi.org/10.1016/j.cirpj.2011.03.007

Seow Y, Goffin N, Rahimifard S, Woolley E (2016) A "design for energy minimization" approach to reduce energy consumption during the manufacturing phase. Energy 109:894–905. https://doi. org/10.1016/j.energy.2016.05.099

Seuring S (2013) A review of modeling approaches for sustainable supply chain management. Decis Support Syst 54(4):1513–1520. https://doi.org/10.1016/j.dss.2012.05.053

Seuring S, Müller M (2008) From a literature review to a conceptual framework for sustainable supply chain management. J Clean Prod 16(15):1699–1710. https://doi.org/10.1016/j.jclepro. 2008.04.020

Simon H (1977) The new science of management decision. Prentice Hall, Upper Saddle River

Solding P, Petku D, Mardan N (2009) Using simulation for more sustainable production systems – methodologies and case studies. Int J Sustain Eng 2(2):111–122. https://doi.org/10.1080/193 97030902960994

Sproedt A, Plehn J (2012) Environmental value stream map as a communicative model for discrete-event material flow simulation. In: 4th world P&OM international annual EurOMA conference 2012

Sproedt A, Plehn J, Hertz P (2013) A simulation enabled procedure for eco-efficiency optimization in production systems. In: Advances in production management systems. Sustainable production and service supply chains, pp 118–125

Sproedt A, Plehn J, Schönsleben P, Herrmann C (2015) A simulation-based decision support for eco-efficiency improvements in production systems. J Clean Prod 105:389–405. https://doi.org/ 10.1016/j.jclepro.2014.12.082

Stahl B, Taisch M, Cannata A et al (2013) Combined energy, material and building simulation for green factory planning. In: Re-engineering manufacturing for sustainability. Springer Singapore, Singapore, pp 493–498

Stiel F, Michel T, Teuteberg F (2016) Enhancing manufacturing and transportation decision support systems with LCA add-ins. J Clean Prod 110:85–98. https://doi.org/10.1016/j.jclepro.2015. 07.140

Stoldt J, Putz M (2017) Procedure model for efficient simulation studies which consider the flows of materials and energy simultaneously. Procedia CIRP 61:122–127. https://doi.org/10.1016/j. procir.2016.11.195

Thiede S (2012) Energy efficiency in manufacturing systems. Springer, Berlin, Heidelberg

Thiede S, Li W, Kara S, Herrmann C (2016a) Integrated analysis of energy, material and time flows in manufacturing systems. Procedia CIRP 48:200–205. https://doi.org/10.1016/j.procir. 2016.03.248

Thiede S, Schönemann M, Kurle D, Herrmann C (2016b) Multi-level simulation in manufacturing companies: the water-energy nexus case. J Clean Prod 139:1118–1127. https://doi.org/10.1016/ j.jclepro.2016.08.144

Thies C, Kieckhäfer K, Spengler TS, Sodhi MS (2019) Operations research for sustainability assessment of products: a review. Eur J Oper Res 274(1):1–21. https://doi.org/10.1016/j.ejor.2018.04.039

Thollander P (2008) Towards increased energy efficiency in Swedish industry – barriers, driving forces & policies. Linköping University

Thollander P, Mardan N, Karlsson M (2009) Optimization as investment decision support in a Swedish medium-sized iron foundry – a move beyond traditional energy auditing. Appl Energy 86(4):433–440. https://doi.org/10.1016/j.apenergy.2008.08.012

Ude J (2010) Entscheidungsunterstützung für die Konfiguration globaler Wertschöpfungsnetzwerke. Shaker Verlag, Aachen

United Nations Environment Programme (2009) Guidelines for social life cycle assessment of products. Nairobi

van der Vorst J, Tromp S-O, van der Zee D-J (2009) Simulation modelling for food supply chain redesign; integrated decision making on product quality, sustainability and logistics. Int J Prod Res 47(23):6611–6631. https://doi.org/10.1080/00207540802356747

Verl A, Westkämper E, Abele E et al (2011) Architecture for multilevel monitoring and control of energy consumption. In: Hesselbach J, Herrmann C (eds) Glocalized solutions for sustainability in manufacturing. Springer Berlin Heidelberg, Berlin, Heidelberg, pp 347–352

Weinert N, Fischer J, Posselt G, Herrrmann C (2013) Lean and green framework for energy efficiency improvements in manufacturing. In: Proceedings of the 11th global conference on sustainable manufacturing – innovative solutions, pp 461–466

Weinert N, Fink R, Mose C et al (2014) Comprehensive improvement of industrial energy efficiency: pilot case in a European rolling stock factory. IFIP Adv Inf Commun Technol 439:334–341. https://doi.org/10.1007/978-3-662-44736-9_41

Widok AH, Wohlgemuth V (2016) Integration of a social domain in a manufacturing simulation software. Int J Serv Comput Orient Manuf 2(2):138. https://doi.org/10.1504/IJSCOM.2016.076429

Widok A, Schiemann L, Jahr P, Wohlgemuth V (2012) Achieving sustainability through a combination of LCA and DES integrated in a simulation software for production processes. In: Proceedings of the winter simulation conference 2012, pp 1219–1229

Winkler H (2011) Closed-loop production systems – a sustainable supply chain approach. CIRP J Manuf Sci Technol 4(3):243–246. https://doi.org/10.1016/j.cirpj.2011.05.001

Wohlgemuth V, Widok A (2016) Konzeption und Entwicklung eines betrieblichen Umweltinformationssystems zur Simulation von Aspekten der Nachhaltigkeit bei Produktionssystemen. In: Mokosch M, Urban T (eds) Umweltinformationssysteme: Definition, Bedeutung und Konzeption. Shaker, Aachen, pp 69–79

Wohlgemuth V, Page B, Kreutzer W (2006) Combining discrete event simulation and material flow analysis in a component-based approach to industrial environmental protection. Environ Model Softw 21(11):1607–1617. https://doi.org/10.1016/j.envsoft.2006.05.015

Wohlgemuth V, Schnackenbeck T, Panic D, Mäusbacher M (2008) Development of an open source software framework as a basis for implementing plugin-based environmental management information systems (EMIS). In: International symposium on environmental software systems, pp 584–592

Zein A (2013) Transition towards energy efficient machine tools. Springer, Berlin, Heidelberg

Zhou M, Pan Y, Chen Z, Yang W (2013) Optimizing green production strategies: an integrated approach. Comput Ind Eng 65(3):517–528. https://doi.org/10.1016/j.cie.2013.02.020

Chapter 4
Concept to Increase Resource Efficiency in Manufacturing Value Chains

In this chapter, a new concept to increase resource efficiency in manufacturing value chains is presented. Section 4.1 deals with the translation of research demands into specific requirements for a new concept. In Sect. 4.2, an overview about the new concept and its inherent phases and steps is provided, followed by a detailed description of its three consecutive phases in Sects. 4.3, 4.4 and 4.5. Finally, in Sect. 4.6 implications for an operationalization of the concept into a *decision support toolbox* are summarized.

4.1 Requirements

Requirements have been derived based on the research questions as presented in Sect. 1.2. They refer to specific characteristics of the new concept to answer these questions. They also aim to close the gaps identified in existing work (compare Sect. 3.4) by fulfilling the evaluation criteria derived in Sect. 3.2. The concept is developed in a deductive way, starting with rather general requirements in order to meet the needs of various industrial actors to increase their resource efficiency. Models for specific manufacturing systems will be presented later for exemplary industrial use cases in Chap. 6. Table 4.1 provides an overview and a short description of all requirements dealing with research question 1, i.e. with methods to assess resource efficiency in manufacturing value chains.

Regarding the second research question, further requirements are derived as indicated in Table 4.2. Referring to the evaluation criteria described in Sect. 3.2, they cover both *decision support* and *application*.

© The Author(s), under exclusive license
to Springer Nature Switzerland AG 2020
S. A. Blume, *Resource Efficiency in Manufacturing Value Chains*,
Sustainable Production, Life Cycle Engineering and Management,
https://doi.org/10.1007/978-3-030-51894-3_4

Table 4.1 Specific requirements addressing research question 1

RQ1	How to holistically assess resource efficiency in manufacturing value chains?
R1	**Cradle to grave modeling from product perspective**: The concept shall facilitate a consideration of the entire product life cycle from cradle to grave, i.e. from raw material extraction until end of life stage. This scope is required to avoid problem shifting between different product life cycle stages (compare Sect. 1.1). Still, the main focus of modeling shall be put on the production stage, as actors from the manufacturing sector are intended to be the potential target group of the concept. Models with a scope from cradle to grave shall allow them to judge about the global consequences of their manufacturing decisions.
R2	**Consideration of multiple manufacturing system levels from machine level up to value chain level**: The concept shall be suited to provide decision support for stakeholders on different levels such as production planners, factory planners, energy managers, environmental managers, managing directors, supply chain managers etc. Consequently, resource flows, performance indicators (PIs) and improvement measures on different levels of manufacturing systems from process level up to value chain level shall be taken into account (compare Sect. 2.1.2). Consequences of local decision making (e.g. changes implemented on machine level) on higher system levels and vice versa shall become transparent, while potential advantages and drawbacks of actor-spanning cooperation shall become quantifiable.
R3	**Consideration of all relevant resource demands**: The new concept shall be qualified for making well-grounded manufacturing decisions. As human labor, materials and energy are on average the main cost driving resources in manufacturing (compare Sect. 2.1.1), they shall be considered during modeling and decision making. This shall be achieved by creating detailed material and energy flow models (compare MEFA method in Sect. 2.4.2).
R4	**Multi-criteria analysis of manufacturing systems**: Motivated by the manifold management objectives pursued by producing companies, a broad range of PIs shall be considered to cover different target dimensions (compare Sect. 2.1.3). Beneath technical PIs (e.g. lead time, resource utilization), also economic PIs (e.g. production costs) and environmental PIs (e.g. global warming potential) are of interest. In contrast, a social assessment is not aspired due to the significant additional effort regarding data collection. Existing and well-established modeling and evaluation methods typically address only one of these dimensions (compare Sect. 2.4). In this concept, they shall be combined and adapted to achieve a more holistic manufacturing system evaluation. Due to a common data basis used for their application, integrity of the results should be increased compared to existing, independently applied methods.
R5	**Support in metering and data interpretation**: Data collection and interpretation as prerequisites for system modeling often go along with significant efforts and costs. As an example, high amounts of energy metering data typically need to be processed and condensed in order to support decision making in terms of energy efficiency improvements (Posselt 2016). The proposed concept shall reduce efforts and dependency on experts for data interpretation. This shall be achieved by applying adequate methods, which assist humans in extracting useful information from big volumes of data.
R6	**Verification and validation**: Models are abstracted representations of real-world systems, i.e. simplifications need to be made without significantly affecting the general system behavior (compare Sect. 2.4). Thus, a verification of calculations is mandatory to ensure their correctness. Further, a validation shall be conducted by applying the concept to real industrial use cases.

Table 4.2 Specific requirements addressing research question 2

RQ2	How to provide decision support for system improvement, assuming limited resources such as time and expert knowledge?
R7	**Systematic hot spot analysis**: The challenge to find the right spot for measure application as described in the research demand in Sect. 3.4 shall be tackled by using appropriate methods. They shall allow to determine the impact of single system parameters on overall PIs. Therefore, multiple parameter variations need to be conducted to identify the most influencing parameters and provide a first prioritization. As results, main drivers for the overall system performance in terms of costs or environmental impacts shall be identified.
R8	**Benchmarking of process performance against reference values**: In order to close the existing gap regarding an adequate evaluation of process efficiencies, a benchmarking against reference values shall be integrated into the new concept. Both empirical reference values but also benchmarks build on physical models shall be used to identify processes with high saving potentials.
R9	**Support in measure identification**: As the dependency on expert knowledge to identify improvement measures is typically high, support in this crucial step shall be provided. Therefore, a knowledge-based system (KBS) shall be developed to support human actors in decision situations (compare Sect. 2.3.4). They can make available existing knowledge and provide recommendations for improvements. As a result, the reliance on experts shall be reduced while a high quality of decisions is maintained.
R10	**Prediction of improvement measure impacts**: This requirement refers to the ability of assessing both the initial (as-is) state but also alternative (what-if) scenarios. The new concept shall allow to predict consequences of manufacturing decisions such as technical and organizational improvement measures. The focus shall be put on tactical and strategic decision with relation to energy and material flows (e.g. implementation of new technologies, change of resource suppliers or characteristics, closing of material loops).
R11	**Ranking of improvement measures in accordance with individual targets**: A systematic ranking of evaluated improvement measures shall be facilitated in order to support decision making. Therefore, methods for multi-criteria decision analysis shall be integrated into the concept (compare Sect. 2.3.3). They allow to rank alternatives under consideration of individual targets and potential target conflicts between different evaluation criteria.
R12	**Comprehensible application procedure**: The provision of an application procedure shall allow to use the developed concept in industrial practice in a meaningful way and with reproducible results. Therefore, the different steps during modeling, evaluation and improvement of manufacturing systems need to be thoroughly described. Critical steps need to be highlighted and recommendations for critical decisions need to be given, also highlighting consequences of alternative courses of action.

(continued)

The requirements presented ahead are taken into account for concept development as described subsequently. Their fulfilment is addressed during the description of concept steps in Sects. 4.3–4.5.

Table 4.2 (continued)

RQ2	How to provide decision support for system improvement, assuming limited resources such as time and expert knowledge?
R13	**User-friendly software implementation featuring a high degree of automation**: The concept application shall be facilitated by the provision of a software tool, whereby only one model shall be used as basis for the analyses. Accordingly, the challenge is to integrate individual but complementary methods as described in Sect. 2.4 into one software tool. Further features of the software implementation shall comprise a comprehensible visualization of results, a consistent data management as well as a consistent calculation of PIs. Further, a high maturity of the software tools is aspired. A good usability is regarded as a basic paradigm for tool development in order to facilitate decision making (Turban et al. 2007). Thus, a standard software shall be used instead of expert software tools in order to achieve a time and cost-efficient application. This may also foster the dissemination of results, as models could be opened and adapted from almost every workstation.
R14	**Pre-defined model elements to reduce modeling efforts**: Modeling complex manufacturing value chains is usually related to high modeling efforts. Following the idea of some existing approaches, pre-defined models for main system elements shall be developed and saved to a model gallery. They can then be used during modeling, significantly reducing the efforts compared to usual modeling from scratch. At the same time, the probability of errors during modeling can be reduced by verifying the correctness of pre-defined elements.
R15	**Scalability of approach to cover use cases with varying level of detail and complexity**: The proposed concept shall target at a holistic evaluation of resource efficiency in manufacturing. This goes along with needs for a detailed system modeling and extensive data acquisition. However, not every use case demands for this high level of detail. Sometimes, only selected subsystems and resources need to be analyzed. Hence, a scalability of the concept shall be given, making it applicable for very broad and detailed cases on the one hand but also for quick and efficient analyses with a limited scope on the other hand.
R16	**Adaptability and extendibility to suit different industrial branches and use cases**: The concept shall be developed for general applicability in manufacturing industries. Due to the individuality of companies and the great variety of potential products and processes in industry, the need to integrate new system elements or resources may occur in the future. Consequently, concept and tool shall be adaptable, e.g. by allowing to customize pre-defined model elements.

4.2 Concept Overview

A new concept to improve resource efficiency in manufacturing value chains is developed based on the research demands and requirements discussed beforehand. It intends to evaluate the technical, economic and environmental performance of manufacturing systems producing discrete products by analyzing related resource flows. The concept promotes the abstraction of real-world value chains into virtual system models from a product perspective as depicted in Fig. 4.1. These models can cover the whole product life cycle (raw materials, production, use, end of life), while the main focus is put on the production stage. Thus, modeling this stage is mandatory and features a high degree of detail, while a consideration of other production stages

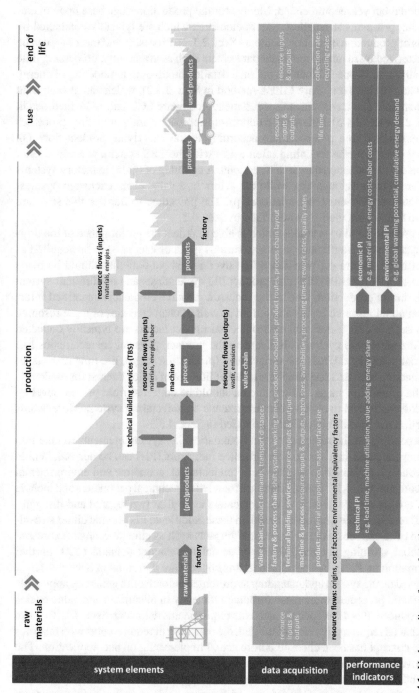

Fig. 4.1 Understanding of manufacturing value chains applied for concept development

is optional but yet recommended. Models for the production stage base upon manufacturing processes as smallest system elements, which are typically conducted by means of production machines (compare Sect. 2.1.2). Processes and machines can be characterized by manifold technical parameters such as availability, processing time or quality rate. A special focus is put on a detailed input–output modeling of energy and material flows (compare MEFA method in Sect. 2.4.2), which can also be used for economic and environmental assessments (compare LCC and LCA methods in Sects. 2.4.4 and 2.4.5). Following a bottom-up approach during modeling, processes can be coupled to a process chain according to the underlying product flow. On factory level, a further coupling often exists with the TBS system, whereas energy and media are exchanged in both directions. A coupling of multiple factory systems is referred to as value chain, while the actors in a value chain exchange physical products (supplier-customer-relationship). The procedure to transfer this structure into virtual system models is described in Sect. 4.3.

In order to build up a model which is able to reflect the performance of the original system with a sufficient accuracy, various types of data need to be acquired as model input. Figure 4.1 provides a first overview about data that should be made available with regard to different product life cycle stages and on different system levels during production. Due to the focus on resource efficiency, quantified information about resource inputs and output as well as characteristics of these resource flows such as cost factors and environmental impact factors are typically collected over the whole life cycle. With regard to the product, physical characteristics like material composition, mass or surface size (production stage), life time (use stage) or collection rates and recycling rates (end of life) are of major interest for modeling. For the production stage, additional data should be collected from process level up to value chain level in order to characterize the manufacturing system. More details and guidance on data collection is provided in Sect. 4.3.2.

Depending on the selected system boundaries, i.e. system elements to take into account for modeling, various performance indicators (PIs) can be derived. Within this concept, a distinction is made between technical, economic and environmental PIs (compare Fig. 4.1). Technical PIs are founded on technical parameters and include time, quality and flexibility related indicators as well as (aggregated and disaggregated) resource flow quantities. Based on these, additional PIs provide either an evaluation of related costs or environmental impacts such as climate change (expressed as global warming potential *GWP*) or cumulative energy demand *CED*. Further information on PIs to be considered in this concept are presented in Sect. 4.4.1.

Based on the system understanding as presented beforehand a generally applicable systematic procedure to increase resource efficiency in manufacturing value chains is introduced. It is developed to match the requirements defined in Sect. 4.1. To make sure that all requirements can be fulfilled, the matching of requirements with modules of the concept has been checked before actual implementation and application. The improvement procedure covers all steps to model, evaluate and improve single factories and entire value chains. It is oriented towards typical problem solving cycles such as plan-do-check-act (PDCA) (Moen and Norman 2009) as well as procedures from systems engineering (Daenzer and Huber 1994). Further, existing application procedures of approaches analyzed in the state of research have been taken into account such as Mardan and Klahr (2012), Thiede (2012) and Heinemann (2016). During the development, a focus is put on the identification of methods and elements to provide decision support in all steps of the procedure. Existing and well-established solutions (e.g. from approaches presented in Sect. 3.3.2) are taken into account. The iterative improvement procedure as depicted in Fig. 4.2 comprises three phases – modeling, evaluation and improvement. In each phase, different procedure steps and underlying methods are proposed:

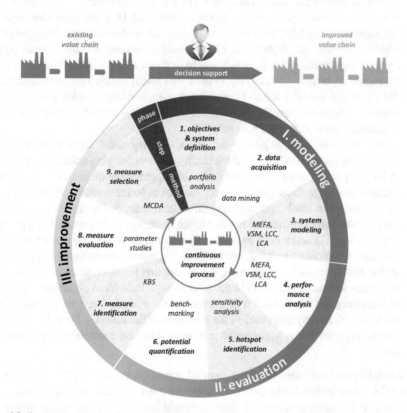

Fig. 4.2 Improvement procedure containing three phases with inherent procedure steps and selected underlying methods

1. The procedure starts with the *modeling phase*, containing *objectives and system definition, data acquisition* and *system modeling*. Depending on the objectives of the analysis, different aspects of the real-world system can be modelled with varying level of detail. Accordingly, type and extent of needed data as well as modeling efforts are highly individual for each use case. Underlying methods comprise portfolio analyses, data mining methods to support data acquisition and MEFA as basis to create system models based on material and energy flows (compare Sect. 2.4.2).

2. In the subsequent *evaluation phase*, a *performance analysis* is carried out to assess the initial state of the value chain in terms of resource efficiency. VSM is applied to assess the technical performance (compare Sect. 2.4.1), while economic aspects are assessed using LCC (compare Sect. 2.4). LCA is applied to gain transparency about environmental impacts related to resource flows over the whole product life cycle (compare Sect. 2.4.5). A *hot spot identification* is performed in a next step by means of sensitivity analyses to identify critical parameters, i.e. the drivers for costs and environmental impacts. A *potential quantification* using benchmarking allows to evaluate the performance of single manufacturing processes in comparison to reference processes.

3. In the final *improvement phase*, the *measure identification* aims at an improvement of the system (factory or value chain), e.g. by changing critical parameters. A KBS supports during the identification of improvement measures by providing expert knowledge about promising approaches to increase resource efficiency on different system levels and related to different resources. During *measure evaluation*, identified measures are applied to the system model by conducting parameter studies. Consequences regarding the system performance can be assessed and used for decision making. For the final step *measure selection*, an application of MCDA methods (compare Sect. 2.3.3) supports to select the most meaningful measures for implementation in accordance with individual objectives. As soon as measures are implemented in the real-world system, hotspots and improvement potentials may shift, calling for repetition of the analysis. For this reason, an iterative approach is recommended in order to achieve a continuous improvement process.

Modeling and evaluation methods applied in the concept, i.e. MEFA, VSM, LCC and LCA, are selected due to their complementary strengths as indicated in Fig. 4.3. Benefits of integrating several of these methods have already been demonstrated by Thiede et al. (2016). Evaluation criteria are oriented towards criteria presented in Sect. 3.2. Accordingly, the selected method combination allows for a holistic assessment of manufacturing system from a technical, economic and environmental dimension. The decision to use MEFA as basic modeling paradigm is regarded as prerequisite to apply LCC and LCA. However, these methods do hardly consider technical aspects related to manufacturing system operation. Thus, VSM is used here to assess the technical performance covering time and quality related aspects. Simulation, in contrast, is not included despite its flexibility and its ability to provide a high transparency in terms of (dynamic) technical system performance. This decision

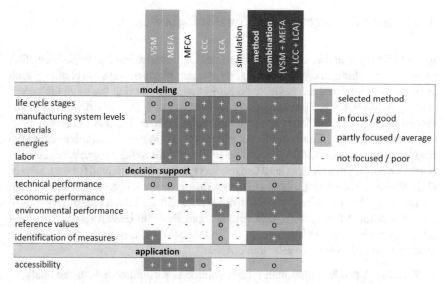

Fig. 4.3 Comparison of methods for system modeling and evaluation as basis for method selection

is made due to keeping the complexity of the resulting concept manageable, as a dynamic simulation usually necessitates the application of specialized simulation software and extensive expert knowledge. This would be contradictory to the aspired high degree of applicability (compare requirement R13). As a drawback, dynamic technical interactions within the manufacturing system cannot be fully analyzed, entailing limitations in operational decision making (e.g. production scheduling).

4.3 Phase I: Modeling

The first phase *modeling* constitutes the basis of the concept as it describes the system models on which all further steps are depending. Within the phase, three consecutive steps can be distinguished: the *objectives and system definition*, setting main objectives, system boundaries and performance indicators for system performance evaluation, is described in Sect. 4.3.1. Guidance on *data acquisition* is given in Sect. 4.3.2, providing both strategies and typical sources for data collection. Further, support in data interpretation is provided applying data mining approaches. They allow to extract useful information (knowledge) from big volumes of low level data (Fayyad et al. 1996). As it is highly important to choose a suitable scope and level of detail for modeling, in Sect. 4.3.3 a particular focus is put on guidance for *system modeling* using the MEFA method (compare Sect. 2.4.2) in accordance with the objectives followed.

4.3.1 Step 1: Objectives and System Definition

In the first step, suitable objectives and system boundaries need to be defined. This early system definition ensures that no relevant manufacturing system elements such as processes and flows are neglected during modeling and that the model is appropriately detailed for the intended purpose. Furthermore, an early definition of the system is a requirement for an efficient data acquisition and allows to identify most suitable data sources. The definition of objectives and system boundaries starts with the selection of a real-world system to be analyzed. Due to the chosen product focus (compare requirement R1), at least one product type must be selected for the analysis. Only elements with direct relation to the product creation will be taken into account for modeling. For traditional VSM (compare Sect. 2.4.1), Erlach (2010) proposes to choose a product family with high volumes and share on overall revenues but not too many variants for an initial analysis. In this concept, it is suggested to select a product type following two selection criteria:

- **Volume**: A product type with a high volume is recommended for the analysis. This can refer to products which are sold in high quantities or which represent a high share of total costs or revenues. Due to the volume, slight increases in the production of a single product may entail significant overall improvements.
- **Performance**: If products with a low performance in terms of economic competitiveness, reject rates etc. are already known, they can be selected in order to further analyze the causes for their underperformance. For products with an insufficient

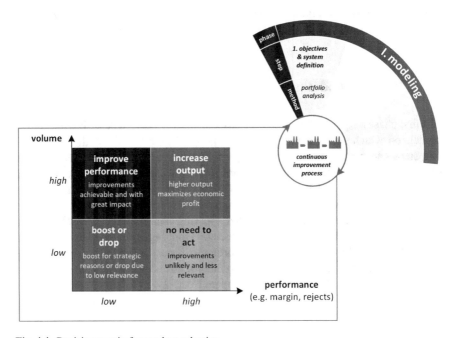

Fig. 4.4 Decision matrix for product selection

margin, an analysis can support decision making about potential outsourcing or even a stop of production.

Figure 4.4 proposes a decision matrix to facilitate the selection of case products to be analyzed. Accordingly, products with high volume and low performance should be selected for evaluation with priority, whereas products with high performance and low volume can usually be neglected. In the remaining two cases, products might for instance be selected to maximize profit by further increasing the volume (high performance, high volume) or to become more competitive by boosting the performance of a product featuring low performance and volume.

The selection of case product types is inseparably linked to the definition of objectives. Almost every analysis aims at gaining transparency and system under-standing, i.e. knowledge about cause-effect relations of system elements as intro-duced in Sect. 2.1.2. Apart from that, an improvement of the system performance in either one or several evaluation dimensions is regarded as basic motivation to apply the concept to a real-world case. System boundaries, i.e. life cycle stages, manu-facturing system levels and considered resource flows shall then be chosen accord-ingly. A general decision support in dependence on different underlying objectives is provided in Fig. 4.5. Depending on the main objective, it proposes to focus modeling on specific life cycle stages, manufacturing system levels as well as resource flows. In terms of considered product life cycle stages, a sole analysis of the production stage can be carried out in the simplest case of a first hot spot identification, if no detailed technical, economic or environmental assessments have been conducted beforehand. Extensions with raw materials stage, use stage and end of life stage can be carried out. This applies in particular for environmental analyses in order to avoid problem shifting along the product life cycle (addresses requirements R1, R16). Within the production stage, the analysis can either focus on one or few key processes, but also comprise complete process chains and TBS within a factory or – in the broadest case – interlinked factories along a value chain (R2, R16). As an example, a focus on technical objectives such as increased production throughput often allows to limit

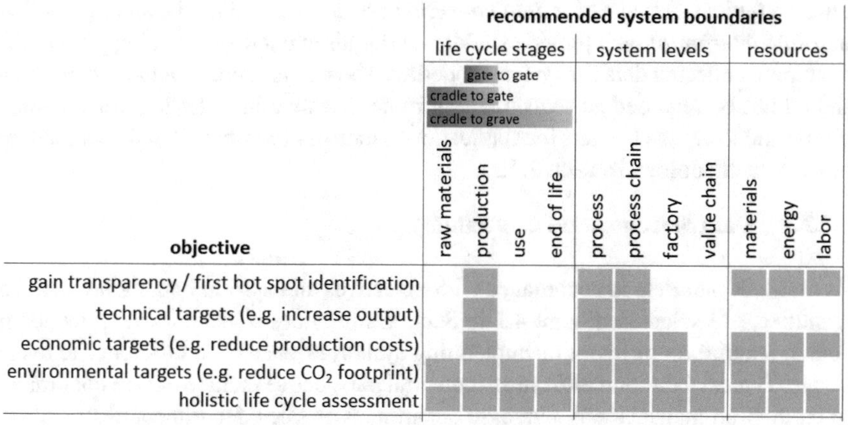

objective	recommended system boundaries										
	life cycle stages				system levels				resources		
	raw materials	production	use	end of life	process	process chain	factory	value chain	materials	energy	labor
gain transparency / first hot spot identification		▪			▪				▪	▪	▪
technical targets (e.g. increase output)		▪			▪	▪			▪	▪	▪
economic targets (e.g. reduce production costs)		▪			▪	▪	▪		▪	▪	▪
environmental targets (e.g. reduce CO_2 footprint)	▪	▪	▪	▪	▪	▪	▪	▪	▪	▪	▪
holistic life cycle assessment	▪	▪	▪	▪	▪	▪	▪	▪	▪	▪	▪

Fig. 4.5 Decision support to select system boundaries in dependence on objectives

the analysis on a process chain or even a single process without taking into account resource flows apart from the actual product flow. A limitation of the analysis to the most relevant resource flows is possible, for instance to raw materials contained in the product, electricity or labor (R3, R16). Resource flows should be selected due to related costs or environmental impacts (if known). The selection can be based on existing analyses. If resources are of negligible relevance, it is recommended to refrain from their consideration to reduce efforts and complexity for data acquisition and modeling.

If only specific improvement measures with a predictable range of effects and related resources shall be assessed, the model can also be tailored very flexibly to the needs, neglecting irrelevant processes and resources.

Example

The replacement of a machine component may only influence the direct electricity demand of that machine. The scope of the analysis can then be limited to the specific process of interest. In contrast, a re-design of products often entails changes in several processes along an entire value chain but also in other life cycle stages (e.g. different materials in raw materials stage, different processing parameters in production stage, different energy demand in use stage, different recycling strategy in end of life stage). Thus, their consideration in modeling is highly recommended.

4.3.2   Step 2: Data Acquisition

The second step of the modeling phase deals with data acquisition. On the one hand, it comprises the activity of *data selection*, identifying relevant measurands. In this context, potential company internal and external data sources are briefly discussed. Further, typical strategies for *data collection* are described. Finally, an introduction into *data interpretation* is provided (addresses requirement R5), including procedures to prepare collected data for system modeling. The whole process of data acquisition should not be regarded as sequential but rather iterative in order to reduce related efforts and lower the barriers for application in industry (compare iterative procedure of LCA methodology in Sect. 2.4.5).

4.3.2.1   Data Selection and Collection

At first, relevant data in accordance with objectives and system boundaries shall be identified and selected. Figure 4.1 in Sect. 4.2 provides a general overview about data to acquire on different manufacturing system levels and in different life cycle stages. Typical data sources from an industrial perspective include but are not limited to production management, purchase department or sales department. As methods suitable for operational and real-time decision making like simulation are not taken into account, there is no general need to acquire all data with a very high time resolution such as seconds or minutes. Instead, cumulative data is often sufficient, representing valid average values for certain product types or time periods. Nevertheless, in some cases a higher timely resolution and therefore further data processing might be needed, e.g. for electrical load profiles of machines (compare Sect. 4.3.3.2).

After selecting the relevant data types in accordance with objectives and system boundaries, the actual data collection can be conducted. Historical data, reflecting the current system behavior, shall be collected in this step to create an initial system model and evaluate the system's initial performance. In *step 8: measure evaluation*, additional data might be required to assess the potential impact of improvement measures. A couple of strategies to collect data are established in practice, depending on objectives and framework conditions as well as the type of collected data (e.g. primary or secondary, qualitative or quantitative) (Sapsford and Jupp 2006; Cooper and Schindler 2014). In this concept, three general options for data collection are proposed: expert interviews, document analyses and measuring & metering. According to their characterization presented in Fig. 4.6, expert interviews provide rather aggregated, personally interpreted and therefore subjective information. In contrast, direct measuring and metering typically features a higher validity and timely resolution of data, but may necessitate further interpretation to derive helpful information. Related efforts and costs often increase accordingly.

The proposed procedure for collecting data starts with a conduction of *expert interviews* in relevant company departments. With their theoretical knowledge and practical experience, experts can provide first insights about relevant system elements and critical system parameters such as cost drivers. Thus, expert interviews help to set priorities for further data collection and can also give indications about relevant data sources for *document analyses*. Depending on the required data, several different documents like technical documentations, product data sheets, shift plans, energy bills or reports from MES and ERP systems may be relevant (Li et al. 2011; Thiede 2012; Despeisse 2013; Posselt 2016). Typically, they are distributed among various stakeholders and a main challenge in practice often relates to the identification of and

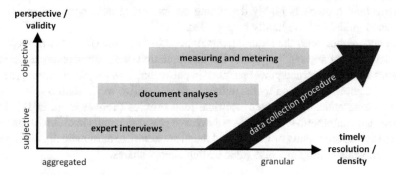

Fig. 4.6 Characterization of different data collection strategies and proposed procedure

collaboration with these stakeholders. Further, documents may significantly differ regarding actuality and interpretability, demanding for a thorough analysis. In order to close data gaps and to validate data collected from interviews and document analyses, direct *measuring and metering* can be applied. In some cases, a sole visual observation of processes or objects may be sufficient, e.g. in order to understand product flow routes (Erlach 2010). However, in most cases specific metering equipment is needed and related efforts are significant. A thorough planning of metering is essential in any case, including the definition of an appropriate metering strategy. In the context of energy metering, Posselt (2016) identified several factors to take into account for deriving metering strategies. They comprise targeted data accuracy, validity, granularity, existing production infrastructure and metering infrastructure.

Depending on the resource flow and data type to be measured, appropriate metering strategies, principles and suitable metering devices differ significantly. Examples with main focus on energy can be found in Herrmann et al. (2010), Kara et al. (2011) and Posselt (2016). For energy metering, Posselt (2016) proposes four consecutive steps, starting with estimation over one time measurements with mobile devices until continuous sampling using permanently installed metering devices, which permanently feed data into IT systems.

4.3.2.2   Data Interpretation

Data interpretation is regarded as a subsequent activity to data collection within the presented concept. It refers to the need for processing collected data before it can actually be used as input for modeling. Data interpretation can be described as a step of contextualization, creating information from data to increase its usefulness (Ackoff 1989). Relevant data might be filtered, aggregated, converted or combined. In practice, data interpretation is often related to high manual efforts, particularly if large quantities of metered data must be processed. In the context of resource efficient manufacturing, this applies in particular to metered energy data: energy demands of machines or TBS are usually dynamic and therefore typically recorded with a high timely resolution (Posselt 2016; Beier 2017). The interpretation of other data than electric load profiles is highly depending on individual data type and source, so universal guidelines can hardly be provided.

In accordance with the requirements defined in Sect. 4.1 (R5), it is proposed to use data mining approaches to ease the interpretation of dynamic resource demand profiles such as electrical power profiles. In particular, *load profile clustering (LPC)*, which has been developed by the author in a cooperative work, shall be introduced here to automatically interpret electrical load profiles (Teiwes et al. 2018). LPC targets at an automatic identification of average power demands in different machine states such as processing or waiting and therefore allows to automatically distinguish between value adding and non-value adding energy shares.

Example

The electrical energy demand of a grinding machine is usually highly state-dependent (Gutowski et al. 2006; Li et al. 2011; Zein 2013). Thus, a distinction between the states *off, standby, operational, working* and *powering up/down* may be adequate in order to assess its energy efficiency (Verband Deutscher Maschinen- und Anlagenbau e.V. 2015). Inspired by this distinction, a classification into the states *off, standby, waiting* and *processing* is applied here, while ramp up and shut down times are no further considered. During non-production times, the machine shown in Fig. 4.7 is in *off* state, not demanding for electricity. During break times, it is operated in *standby* mode, whereby only the machine controls are switched on and demand for ~0.2 kW. In the rest of production time, the machine is either *processing* (average power 3.4 kW) or *waiting* (average power 2 kW). The load during waiting state is also referred to as *basic power* or *fixed power*, which is determined by several components needed to maintain the machine ready for processing (e.g. controls, pumps). Energy demand caused by these components, which do not directly contribute to product modification, is often referred to as *basic energy* (Gutowski et al. 2006; Devoldere et al. 2007; Li et al. 2011). The duration of *waiting* and *processing* depends on the *utilization rate* of the corresponding machine (compare Sect. 4.4.1). Only energy demands in *processing* state are regarded as value adding here, including the corresponding share of *basic energy* in this machine state.

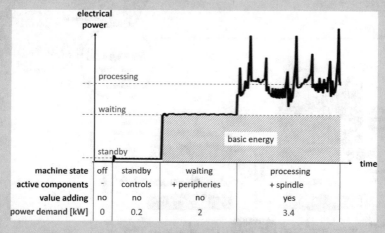

machine state	off	standby	waiting	processing
active components	-	controls	+ peripheries	+ spindle
value adding	no	no	no	yes
power demand [kW]	0	0.2	2	3.4

Fig. 4.7 Exemplary electrical power profile of a grinding machine covering different machine states

The LPC methodology aims at an automatic detection of the machine state *processing* and can therefore distinguish this state from non-value adding machine operations. The methodology comprises five software-supported steps (compare

Fig. 4.8), starting with the actual metering of a demand profile. In step two, a k-means clustering is performed to assign all metered demand values to one of several clusters.[1] Mathematically, the algorithm minimizes the variance J within each cluster according to Eq. (4.1), where x_j refers to the metered values, μ_i to cluster average values and S to the clusters.

$$\min J = \sum_{i=1}^{k} \sum_{x_j \in S_i} ||x_j - \mu_i||^2 \tag{4.1}$$

In a third step, the clusters are assigned to a value adding or non-value adding state of the machine to identify intervals of processing. A matching of these value adding processing intervals with known processing times of the products of interest allows to identify processing periods in the profile. In the last step, a plausibility check is performed. The procedure is either terminated or continued until a plausible processing interval is identified by re-assigning clusters to machine states or performing a re-clustering.

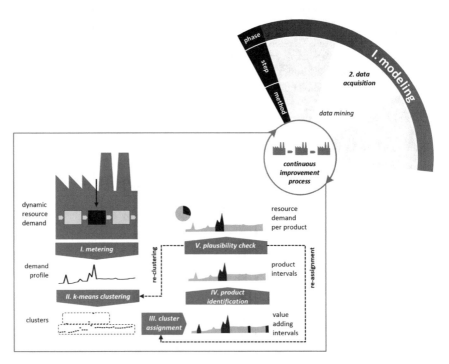

Fig. 4.8 LPC methodology for automatic calculation of product specific energy intensity from load profiles, adapted from Teiwes et al. (2018)

---

[1] K-means is a method of vector quantification, which is often used for cluster analysis (Steinhaus 1956). An overview about its applications in the context of electrical power profile analysis can be found in Teiwes et al. (2018).

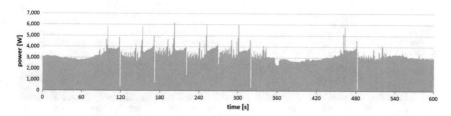

Fig. 4.9 Metered electrical power profile of a milling process during production time

LPC is in principle applicable to metered profiles for several dynamic resource demands on process level (e.g. electricity, compressed air, water) and has therefore the potential to reduce efforts and required knowledge for data interpretation. Here, the application of the LPC methodology shall be demonstrated for electricity with a brief example of a milling process that is conducted with a CNC machining center. An excerpt of the metered electrical power profile, which has been recorded during production time, is shown in Fig. 4.9. From the diagram, a typical load between ~2.8 and ~4 kW can be visually read off with few short-term peaks up to ~6 kW. However, without further knowledge about machine operation during the recorded time period, an assignment of values to processing and other machine states is hardly possible.

By means of the LPC methodology, an automatic identification of processing times and hence a load level calculation can be conducted with high accuracy. In the first step, a k-means clustering is carried out with $k = 2$ (compare Fig. 4.10), i.e. two load cluster are distinguished. Accordingly, every load value is assigned to either cluster 1 (light grey) or cluster 2 (dark grey).

The LPC methodology bases on the assumption that load levels in processing state are higher than in other machine states. Accordingly, in this example load values of cluster 1 are assigned to processing state while values of cluster 2 are assigned to other non-value adding states. In order to check if this clustering does already identify the processing state, the typical processing time for the produced part is needed as supplementary information. In this case, a processing time of ~50 s is known. Considering this information, a self-programmed algorithm automatically seeks for continuous processing intervals that are close to 50 s. As result, an interval of 25 s is identified as best match and the average power during processing is calculated as 3.875 kW (compare Fig. 4.11).

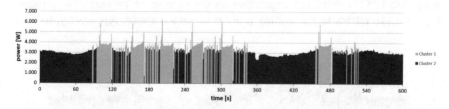

Fig. 4.10 First iteration of clustering load values for a milling process ($k = 2$)

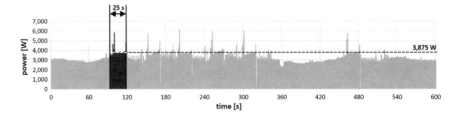

Fig. 4.11 First iteration of product identification for a milling process ($k = 2$)

As shown in Fig. 4.8, a plausibility check needs to be conducted by the user in order to either end the analysis or continue with a re-clustering or re-assignment of cluster values. In this case, a re-clustering is regarded as meaningful due to the big gap between expected processing time (50 s) and the processing interval identified by LPC (25 s). Figure 4.12 shows the result for a re-clustering with $k = 3$. The distinction between three clusters now allows for two alternative assignments: either only *cluster 1* or both *clusters 1 and 2* represent the machine's processing state. According to this logic, the number of possible options to assign processing state and non-value adding states can be expressed as *k-1*. The algorithm implemented in the software tool automatically performs a re-assignment of cluster value to machine states, testing all possible options in order to find the best match in terms of processing time deviation.

The introduction of a third load cluster allows to identify potential processing periods that better match the given processing time of 50 s. The best match can be received, if *cluster 1* and *cluster 2* are both assigned to the processing state. The result is highlighted in Fig. 4.13, indicating a processing interval of 49 s and an average load during processing of 3.448 kW. Due to the low deviation between given and calculated processing time, the received result is taken as final result provided by LPC for this exemplary case.

In order to generally validate the accuracy of results calculated with the LPC algorithm, tests with various metered load profiles are conducted. Each profile is in parallel analyzed manually to achieve a very exact allocation of load values to machine states as reference values. An excerpt of the results is presented in Fig. 4.14, covering four load profiles of different processes. For each corresponding machine, the deviation between exact average load (manual assignment) and calculated average load with LPC is displayed depending on the number of clusters distinguished. Figure 4.14a relates to the milling process presented ahead. Apparently, the first

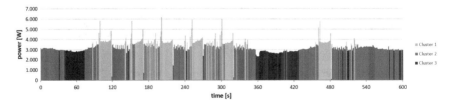

Fig. 4.12 Second iteration of clustering load values for a milling process $k = 3$)

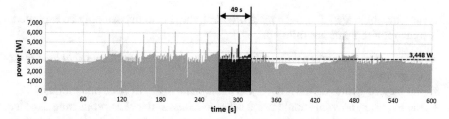

Fig. 4.13 Second iteration of product identification for a milling process ($k = 3$)

distinction of two clusters results in a deviation of 8.3% regarding processing load and 5.8% regarding non-value adding load. These results improved after using $k = 3$, but no further improvement could be achieved for $k = 4$. In analogy, the profiles in Fig. 4.14b–d reveal a generally good accuracy with typical deviations in a single-digit percentage range. However, for profile Fig. 4.14c representing an inductive hardening process, relative deviations in waiting state are very high. This is due to the great difference between load levels for this machine. Few wrongly assigned values then exert significant influence on the lower load level average, i.e. for the waiting state. As a basic rule, a higher number of clusters appears to be generally beneficial to reach more accurate results. As a drawback, calculation times increase exponentially for the distinction of more clusters.

The application of LPC to a couple of metered load profiles revealed strengths and weaknesses of the methodology. The main benefits can be seen in a quick and simple interpretation of load profiles including an automatic calculation of electrical load levels with satisfying accuracy. These can for instance be used as input for the

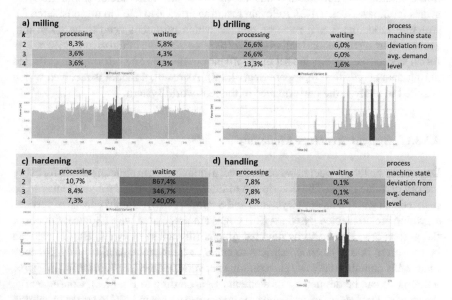

Fig. 4.14 LPC results for exemplary load profiles of four different processes and deviations between exact and calculated resource demand levels

system model in *step 3: system modeling*. Yet, these benefits can only be exploited, if some requirements are fulfilled:

- a clear *relation between value adding machine operation and electrical load* must be given
- the metering interval must cover the *manufacturing of at least one product*
- the metering interval should cover *other machine states than processing* in order to allow the algorithm to differentiate between value adding and non-value adding states
- the metering should be conducted with a *high timely resolution* (e.g. one value per second) in order to preserve load dynamics in the metering data
- the *processing time PT* must be provided to identify a product in the load profile

If these requirements are fulfilled, LPC can serve as helpful decision support for the analysis of large amounts of energy metering data (or other dynamic resource demand data) with low effort and without a need for detailed process knowledge.

4.3.3   Step 3: System Modeling

The third step to be performed refers to the actual buildup of a system model, which mainly bases on material and energy flows (compare MEFA method in Sect. 2.4.2). A bottom-up approach is chosen for modeling due to the intended application of VSM and LCA method, which rely on detailed data on process level. The first activity within the system modeling is related to *system structure modeling*, i.e. to set up the general structure according to the relations of system elements, while the product flow constitutes the horizontal connection between processes, factories and life cycle stages (compare Sect. 2.1.2). Relevant resource flows also need to be considered for structure modeling, i.e. it must be defined which resources need to be taken into account. During the second activity *model parametrization*, initial quantitative values for all system parameters are set in accordance to the data and information collected during data acquisition (compare Sect. 4.3.2). As result, a model is received which is suited to analyze the performance of the real-world system.

4.3.3.1   System Structure Modeling

Depending on the product flows considered, a unique system structure arises for every factory or value chain. The description of interlinked manufacturing systems can be realized using graph theory (Dyckhoff and Spengler 2010). Graphs consist of vertices V, which are connected by edges E. If edges are associated with a logical direction, the graphs are referred to as directed graphs (digraphs) $G = (V, E)$. Graphs can be converted into adjacency matrices to enhance their processability. An adjacency matrix AM contains one row i and column j per vertex of the graph, resulting in an $n \times n$ matrix. It can be defined by its elements a_{ij} according to Eq. (4.2). Consequently, the matrix elements indicate, if either an edge from vertex in row i to vertex in column

j exists (entry 1) or does not exist (entry 0).

$$a_{ij} = \begin{cases} 1 & \text{for } (i, j) \in E \\ \text{else} & 0 \end{cases} \tag{4.2}$$

Adjacency matrices can be used to describe the general structure of processes within a factory (*factory structure matrix*), interlinkage of factories within a value chain (*value chain structure matrix*) and interlinkage of life cycle stages over a product's life cycle (*life cycle structure matrix*). The product flow is therefore represented by the edges, while the vertices are representing manufacturing processes, factories or life cycle stages. Exemplary matrices are presented in Fig. 4.15. Apparently, product flows within a factory can be linear but also converging, diverging or circular (compare Sect. 2.1.2). Causes for non-linear product flows are for instance assembly and disassembly processes, but also defective parts which require reworking in a previous process as well as recycling and reuse of waste materials. For reasons of simplification, product flows along value chains and product life cycle are regarded to be sequential in this concept, hence circular flows are only considered within factory gates.[2]

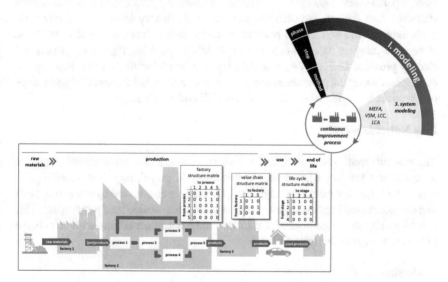

Fig. 4.15 Exemplary system structure with adjacency matrices for factory, value chain and life cycle

---

[2]This simplification is made to keep modeling of manufacturing systems manageable also within the developed tool (compare implementation in Chap. 5). In practice, backflows along the value chain or the product life cycle may occur as well, some of them with a great timely delay during or at the end of the product's life time. In this context, the field of *reverse logistics* is dealing with goods coming back to production from the customer after use for reasons of recycling, remanufacturing etc. (Luger 2010).

In order to ease system modeling, analysis and communication of results, two categories of products are distinguished in system structure modeling. *Factory products* refer to the product outcome of each factory to be handed to the next downstream factory. *Final products* refer to the product outcome of the last factory and thus to the final product of the entire value chain, which is then typically supplied to the customer for use. From a local perspective it is often useful to calculate PIs related to a factory product in order to allow a comparison with other internal analyses. From a global perspective, PIs are rather meaningful to be related to final products. This allows for a better performance comparison between factories, a hotspot detection along the value chain as well as a holistic assessment of improvement measures (compare Sect. 4.5.2). As an example, the energy intensity as PI to express energy demanded per product may differ significantly depending on the chosen reference product.

4.3.3.2 Model Parameterization

After building up the system structure, initial values need to be set for all defined system parameters. This applies for resource flows and parameters on process level including cost factors, but also for parameters on factory level and for other life cycle stages. This first model parameterization shall reflect the initial state of the system, building on data acquired in *step 2: data acquisition*. Figure 4.1 in Sect. 4.2 already provided indications about data types to be collected. In the following, the corresponding model parameters are described in more detail following the sequence of a product life cycle from raw materials until end of life stage.

Raw Materials Stage

The raw materials stage often has a high economic and environmental relevance in terms of a product's life cycle. In this concept, no primary data acquisition is foreseen for the raw materials stage. Still, the stage is considered by allocating costs and environmental impacts to resources supplied from outside the factory gates. This is achieved by an assignment of cost factors end environmental impact factors in the production stage as described below.

Production Stage

As the main focus of the presented concept is put on the production stage, the modeling is conducted with a high level of detail, involving a significant number of system parameters. Figure 4.16 provides an overview about parameters taken into account for modeling of the production stage.

The subsequent explanation of parameters follows a bottom-up approach, starting with products, processes and machines. Regarding the *product* under assessment, Table 4.3 gives an overview about parameters considered.

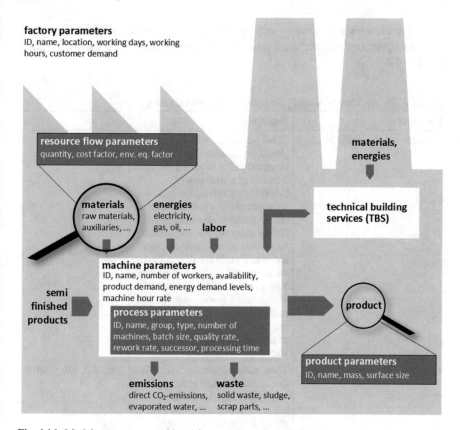

Fig. 4.16 Model parameters considered for modeling of production stage

Table 4.3 Product related model parameters

Parameter	description	Unit
Product ID (ID_p)	This parameter refers to the unique numeric identifier of every factory product p.	–
Product name (N_p)	This parameter refers to the individual name assigned to a product p, allowing for its simpler recognition.	–
Product mass (m_p)	This parameter refers to the physical mass of a product p after the last in-house process. It is mainly required, if PIs are referenced to a product mass, e.g. energy demand per kg of factory product (compare PI specification in Sect. 4.4.1).	kg
Product surface size (A_p)	This parameter refers to the total surface size of a product p. It is mainly required, if PIs are referenced to a product surface, e.g. energy demand per m^2 of factory product (compare definition of PI in Sect. 4.4.1).	m^2

Within a factory, a product typically passes several processes. Regarding physical *resource inputs and outputs*, each process is characterized by the parameters shown in Table 4.4.

Table 4.4 Resource related model parameters considered for each process

Parameter	Description	Unit
Resource input quantity (w, z)	These parameters refer to the physical quantities of resource inputs related to either a quantity of products or a time span of producing a particular product. A differentiation is made between resource inputs from outside the factory gates (w) and intermediate flows from other processes within the same factory (z). Depending on the resource flow, inputs can either refer to a mass (e.g. kilograms of aluminum), a volume (e.g. liters of cooling water), a period of time (e.g. seconds of manual labor) or an amount of energy (e.g. kilowatt-hours of natural gas).	kg, l, s, kWh
Cost factor per input unit (CF_w)	These parameters refer to specific cost factors for each resource input w, which is required for economic calculations (e.g. 0.2 € per kWh electricity). Realistic average buy-in prices should be applied for each resource input.	€/kg, €/l, €/s, €/kWh
Impact factor per input unit (IF_w)	These parameters refer to specific environmental factors for resource inputs w, which are required for environmental calculations (e.g. 0.5 kg CO_{2eq} per kWh of electricity). They are applied for any (raw/auxiliary) material or energy flow w which is supplied from outside the factory gates. Impact factors can be derived from LCA datasets[a] (compare Sect. 2.4.5). The actual source (country or world region) of resources used within a factory should be taken into account for dataset selection.	e.g. kg CO_{2eq}/kWh

(continued)

Table 4.4   (continued)

Parameter	Description	Unit
Resource output quantity (y, r, z)	These parameters refer to physical quantities of resource outputs in analogy to resource inputs. As indicated in Fig. 4.16, this can relate to either products (y), waste or emissions (r for flows leaving the factory, z for flows to be used by another process within the factory).	kg, l, s, kWh
cost factor per output unit (CF_y, CF_r)	These parameters refer to specific cost factors for each resource output due to required post-processing. For product flows and outputs with a certain economic value such as scrap metal, cost factors are typically negative, indicating a revenue per generated output unit.	€/kg, €/l, €/s, €/kWh
Impact factor per output unit (IF_r)	These parameters refer to specific environmental factors for the treatment of resource outputs in analogy to impact factors per input unit.	e.g. kg CO_{2eq}/kWh

[a]LCA datasets are developed applying high quality standards and should be representative for typical industrial activities (United Nations Environment Programme 2011). Still, values provided by LCA databases rather represent average market datasets. Thus, it is important to choose datasets based on conservative assumptions in order to avoid underestimating the relevance of a resource flow (Del Duce et al. 2013). If no suitable dataset for an occurring resource input or output and related processes is available, datasets for products or processes with similar characteristics should be selected.

The quantitative relations between resource inputs and outputs can be formalized using the notation introduced in Sect. 2.4.2 and displayed as input–output tables. The sum of all resource inputs of a system or system component j equals the sum of all its outputs, which can be expressed as follows:

$$\sum_{j=1}^{n} w_j + \sum_{i,j=1}^{n} z_{ij} = \sum_{j=1}^{n} y_j + \sum_{j=1}^{n} r_j + \sum_{i,j=1}^{n} z_{ji} \quad \forall i, j \in N \qquad (4.3)$$

with:

i, j system components
w_j resource inputs from outside the system boundaries to j
z_{ij}, z_{ji} intermediate flows between j and another system element i
r_j waste flows from j leaving the system
y_j usable product flows from j leaving the system

Example

Two processes form a simple process chain. In order to receive one kilogram of final product y, several resource inputs w, intermediate flows z and waste flows r occur as displayed in Fig. 4.17.

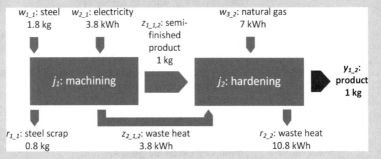

Fig. 4.17 Exemplary resource input and output quantities for simple process chain

Resource inputs and outputs can be displayed as input–output table in order to model flows within the manufacturing system (compare Table 4.5).

Table 4.5   Input–output table summarizing input and output flows of a process chain

		j_1: machining	j_2: hardening	$\sum$
Resource inputs				
w	1: steel (kg)	1.8	0	1.8
	2: electricity (kWh)	3.8	0	3.8
	3: natural gas (kWh)	0	7	7
Intermediate resource inputs and outputs				
z	1: semi-finished product (kg)	-1	1	0
	2: waste heat (kWh)	-3.8	3.8	0
Waste outputs				
r	1: steel scrap (kg)	-0.8	0	-0.8
	2: waste heat (kWh)	0	-10.8	-10.8
Product outputs				
y	1: product (kg)	0	-1	-1

Apart from resource flows, several further *process parameters* are considered on process level (compare Table 4.6). They are mainly derived from traditional value stream mapping (compare Sect. 2.4.1), allowing to calculate the capacity and throughput of manufacturing processes.

The described process parameters are not only relevant for technical analyses, but some do also influence resource input and output quantities and therefore costs and environmental impacts. This influence can be very direct, e.g. if a resource

Table 4.6 Process related model parameters

Parameter	Description	Unit
Process ID (ID_{proc})	This parameter refers to the unique identifier for each process j within a factory. In order to allow for an easier identification of processes along value chains, a combination of factory ID, product ID and process ID is used (e.g. *F1_P3_5* for *process 5*, which is carried out to create *product 3* in *factory 1*).	–
Process name (N_{proc})	This parameter refers to the individual name of each process.	–
Process group (G_{proc}), process type (T_{proc})	These parameters refer to the process group and type in accordance with DIN 8580 (compare Sect. 2.1.2), providing a distinction between six main groups of manufacturing processes, each featuring several sub groups (Deutsches Institut für Normung 2003). The assignment is relevant for results validation, for benchmarking with reference objects (compare Sect. 4.4.3) and for deriving improvement measures (compare Sect. 4.5.1), which are often transferrable among similar processes.	–
Number of machines (NOM_{proc})	This parameter refers to the quantity of machines performing the same process in parallel. This parameter is in particular needed for capacity calculations such as bottleneck analyses (compare interpretation of machine utilization rate *MUR* in Sect. 4.4.1).	-
Batch size (BS_{proc})	This parameter refers to the number of parts that are usually processed by one machine in parallel. Often, it can be counted by observing the process. Information about batch sizes is required for the allocation of resource demands to products. A distinction is made between the number of input parts and output parts per batch, as some processes from main groups separating (e.g. sawing) or joining (e.g. assembly) may change the quantity of parts during the process.	pcs./batch
Quality rate (QR_{proc})	This parameter refers to the share of good parts produced on average. Quality rates below 100 percent result in non-value adding resource demands, which are typically allocated to all good parts.	%
Rework rate (RR_{proc})	This parameter refers to the share of defective parts not directly treated as waste but foreseen for rework. The quantification of a rework rate is only needed for processes with a quality rate of less than 100%	%

(continued)

Table 4.6 (continued)

Parameter	Description	Unit
Successor for rework (SCS_{proc})	If rework is applied to a certain share of parts, a successor for the process must be defined (compare circular flows in Fig. 4.15, Sect. 4.3.3.1).	
Processing time (PT_{proc})	This parameter refers to the average time that a batch of products remains within a process. The PT is in particular relevant for capacity and lead time calculations but it can also be directly related to resource input and output quantities (e.g. a longer processing time may result in a higher energy demand).	s

input quantity depends on the processing time. For electrical energy demands and machining operations, this interrelation almost always applies (Diaz et al. 2009; Zein 2013; Winter et al. 2014; Winter 2016). As explained in Sect. 4.3.2.2, this can be traced back to a typically high fixed power demand of machines compared to the additional variable load induced by actual processing (Gutowski et al. 2006; Zein 2013). Some parameters such as quality rate and rework rate exert indirect influence on resource demands, as they determine the number of products requested from upstream processes or factories. This effect must be taken into account for modeling.

Example

A simple process chain is given as displayed in Fig. 4.18. The influence of quality rate and rework rate of *process 3* on the number of products demanded from upstream processes as well as corresponding resource demands shall be illustrated

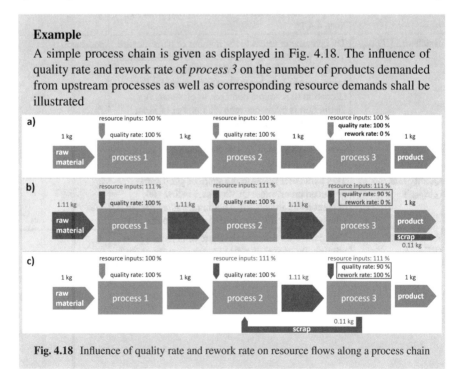

Fig. 4.18 Influence of quality rate and rework rate on resource flows along a process chain

In Fig. 4.18a, the quality rate of *process 3* (and all upstream processes) is assumed to be 100%, thus product input quantities equal product output quantities along the process chain. In Fig. 4.18b, quality problems occur in *process 3* (quality rate *QR* of 90%), i.e. a scrap quantity of 0.11 kg is generated per 1 kg of product due to defective parts. This results in an increased demand for product inputs in all upstream processes, i.e. more raw material or (pre)products need to be processes in each step. Resource demands for energy, auxiliaries, labor etc. for each process rise accordingly with the same factor. As a consequence, *process 1* also demands for a higher raw material input from a previous factory or life cycle stage. Apart from the displayed consequences, other parameters and PIs are affected as well, e.g. the utilization rates of corresponding machines as well as costs and environmental impacts. In Fig. 4.18c, defective parts are not treated as scrap but are instead fully reworked in *process 2* (rework rate *RR* of 100%). Due to the circular material flow between *processes 2* and *3*, no defective part is leaving the process chain as scrap and no higher raw material or other resource demand is induced in *process 1*. However, resource demands of *processes 2* and *3* are identical to Fig. 4.18b. In reality, a reworking of all parts might be unrealistic for most cases, resulting in mixed scenarios of Fig. 4.18b, c.

Every process is carried out using one or multiple *machines*, which are each defined by the parameters listed in Table 4.7.

As processes may be executable on different machines, several process related model parameters are also specific for each process-machine combination. On the one hand, the general usability of a machine for a process needs to be defined. An exemplary process-machine matrix M is shown in Eq. (4.4). Accordingly, *process 1* can be conducted on *machines 1* or *2*, while *process 2* can be conducted on *machines 3* or *4*. *Process 3* can be performed by all machines. Similar matrices can also be created to express the dependence of batch size, quality rate, rework rate and processing time from a machine.

$$M = \begin{array}{c} \\ \text{process} \\ \end{array} \begin{array}{c} \\ 1 \\ 2 \\ 3 \end{array} \overset{\displaystyle \begin{array}{cccc} \text{machine} \\ 1 & 2 & 3 & 4 \end{array}}{\begin{bmatrix} 1 & 1 & 0 & 0 \\ 0 & 0 & 1 & 1 \\ 1 & 1 & 1 & 1 \end{bmatrix}} \tag{4.4}$$

Like indicated in Table 4.8, additional parameters are taken into account on *factory* level in order to define general framework conditions for the manufacturing system.

Table 4.7   Machine related model parameters

Parameter	Description	Unit
Machine ID (ID_{mach})	This parameter refers to the unique numeric identifier assigned to every machine within a factory.	–
Machine name (N_{mach})	This parameter refers to the individual name of each machine within a factory.	–
Number of workers per machine (NOW_{mach})	This parameter refers to the average quantity of workers that are needed to operate the machine. This optional parameter can be used for the calculation of labor inputs, if no more exact data on manual labor demands is available.	–
Availability (AV_{mach})	This parameter refers to the average share of time that a machine is available for value creation during planned working times. Typically, values are below 100% due to machine failures and related need for repair. This parameter is mainly used for capacity analyses.	%
Product demand (PD_{mach})	This parameter refers to the actual number of products which need to be delivered to the next process by a specific machine. As the modeling follows a pull principle according to lean manufacturing (Womack and Jones 1996; Günther and Tempelmeier 2005), the customer demand CD induced by the downstream factory determines the product demands for the whole intra-factory process chain.	pcs./a

(continued)

Table 4.7 (continued)

Parameter	Description	Unit
State-based resource demand levels $(RP_{mach}, RW_{mach}, RS_{mach}, RO_{mach})$	These parameters refer to state dependent resource demands of machines (compare Fig. 4.7 in Sect. 4.3.2.2). The quantification of resource demand levels is optional and typically conducted for relevant energy carriers in order to calculate state dependent energy shares. In particular, a quantification allows for an identification of non-value adding resource demands during non-production times and can therefore help to identify improvement potentials (compare definition of value adding energy intensity in Sect. 4.4.1.1). In this concept, the following machine states and corresponding resource demand levels are distinguished: *Off (RO_{mach})* is assumed for non-production times, i.e. days, shifts and breaks without production activities. Typically, no resources are used in this state. *Standby (RS_{mach})* is applied during break-times according to the current shift system. *Waiting (RW_{mach})* is applied for all times within the official production time that the machine is switched on while it is not processing. *Processing (RP_{mach})* is applied for the individual production time of each machine, which depends on the machine's utilization rate.	e.g. kW
Machine hour rate (HR_{mach})	In this concept, the machine hour rate refers to overhead costs due to machine procurement and operation. Typically this includes depreciations for initial invests as well as costs for maintenance and repair (Götze 2010). Also costs for tool wear or indirect costs for room utilization, lighting, heating, cooling etc. can be subsumed in the machine hour rate, if they are not considered as direct resource inputs of processes or TBS.	€/h

Table 4.8 Factory related model parameters

Parameter	Description	Unit
Factory ID (ID_f)	This parameter refers to the unique numeric identifier of a factory f within a value chain.	–
Factory name (N_f)	This parameter refers to the individual name of each factory, for instance the name of the related company.	–
Factory location (LOC_f)	This parameter refers to the geographic position of a factory. It can be used to calculate transport distances between factories of a value chain and to select suitable LCA datasets for raw materials and end of life stages (compare regionalization of LCA in Sect. 2.4.5).	–
Working days per year (WD_f)	This parameter refers to the average days per year with planned production. For the analysis of multi-product manufacturing systems, only the working days related to the manufacturing of the relevant product type should be considered in order to receive meaningful PIs.	d/a
Working hours per day (WH_f)	This parameter refers to the average hours with planned production per working day. In practice, one, two or three shift systems with approx. eight hours per shift can be typically observed in manufacturing companies. Together with the working days per year, this information is needed for capacity related PI calculation.	h/d
Customer demand (CD_f)	This parameter refers to the quantity of *final products* that is demanded by the next (downstream) factory. The *CD* determines the quantities of products to be produced in each process along a process chain and therefore the product demands per machine *PD*.	products/a

(continued)

Table 4.8 (continued)

Parameter	Description	Unit
Resource input and output quantities of TBS	These parameters refer to resource flows demanded and provided by TBS, which are treated similar to resource inputs/outputs on process level.[a] To allow for a consistent cost and environmental assessment, the demands for intermediate flows generated by TBS such as compressed air need to be broken down into primary resource inputs. For this purpose, average but system specific resource intensity factors RIF are applied. They can for instance be derived from the aggregated resource demand W of the TBS element j and its aggregated provided intermediate flow quantity Z over a certain period as indicated in Eq. (4.5) $$RIF_j = \frac{W_j}{Z_j} \quad (4.5)$$ The resource input w related to a certain intermediate flow z_{ji} supplied by the TBS element j to a process i can then be calculated using Eq. (4.6) $$w_j = RIF_j * z_{ji} \quad (4.6)$$ Resource demands induced by TBS should be handled separately from direct resource inputs on process level, allowing for a more target-oriented identification of suitable improvement measures in *step 7: measure identification.*	kg, l, s, kWh

[a] A separate and detailed modeling of TBS elements is not conducted, as the focus of the concept is put on the actual manufacturing processes. It is assumed, that the process chain under assessment does usually only represent a segment of the factory. However, a sound TBS analysis and improvement would also require to extend the model to a greater part of the production, if TBS infrastructures are shared with other process chains. Although static approaches for TBS evaluation exist, e.g. integrating TBS aspects into energy value stream mapping (Posselt et al. 2014), a thorough TBS analysis typically bases on a dynamic modeling of the manufacturing system (compare Sect. 2.4.6 on simulation).

Use Stage

Depending on the product type, the use stage can account for significant shares of life cycle costs and environmental impacts. As an example, approx. 50–80% of greenhouse gas (GHG) emissions caused over a cars life cycle can be traced back to the use stage (Cerdas et al. 2018). For this reason, well-grounded assumptions should be made regarding the use scenario for the products assessed. The parameters used to describe the use stage in the presented concept are described in Table 4.9.

End of Life Stage

Modeling the end of life stage of products is highly challenging, as processes in end of life are often carried out far in the future. Further, they are often not in direct control of the original product manufacturers. Hence, great uncertainties regarding

Table 4.9 Model parameters considered in the use stage

Parameter	Description	Unit
Product life time (LIT_p)	This parameter refers to the typical period of use for a product before it is passed to the end of life stage. The product life time can either refer to a time period (e.g. use of a smartphone may be four years on average) or other quantifiable parameters (e.g. average car lifetime may be 150,000 km of covered distance). Typical lifetimes for consumer goods can for instance be found in Prakash et al. (2015).	a
Resource input/output quantities (w, r)	These parameters refer to physical quantities of resource inputs and outputs during the use stage from and to other systems. They are treated in the same way as inputs and outputs during production stage. For most products, just a handful of resources like electricity, water or fuel are demanded during use stage. Some products may also cause direct emissions to the environment that should be considered, e.g. products featuring combustion processes.	kg, l, s, kWh
Cost factors per input/output unit (CF_w, CF_r)	These parameters refer to cost factors related with resource demands which need to be defined in analogy to the production stage. However, derived costs are borne by the product user instead of the manufacturer and should be handled separately in *step 4: performance analysis*.	€/kg, €/l, €/s, €/kWh
Impact factors per input/output unit (IF_w, IF_r)	These parameters refer to LCA datasets which relate to selected input and output resources. These are supplied from or released to outside of the factory gates in analogy to the modeling of the production stage.	e.g. kg CO_{2eq}/kWh

future quantity, quality and timing of returned products as well as their processing can be generally stated (Ilgin and Gupta 2010; Allwood 2012). It must be assumed, that many companies do not conduct the end of life treatment of their obsolete products themselves.[3] This situation is referred to as open loop recycling (International Organization for Standardization 2006a). Accordingly, primary data similar to parameters collected for the production stage are so far rarely available, impeding a holistic analysis of products over their life cycles. For these reasons, suitable LCA datasets representing typical end of life treatments for the main fractions of a product should be selected. Losses in both resource quantities and qualities are reflected by assuming typical collection and recycling rates. A summary of parameters used in the end of life stage is given in Table 4.10.

In order to allow for a better understanding of the relations between collection rate, recycling rate and quality degradation rate within this concept, an illustration of considered flows and parameters is provided in Fig. 4.19.

Table 4.10 Model parameters considered in the end of life stage

Parameter	Description	Unit
End of life option ($OP_{p, EOL}$)	This parameter refers to the general end of life (*EOL*) option for the product p such as reuse, remanufacturing, recovery, recycling or disposal (Jawahir et al. 2006). As the complex end of life processes for reuse and remanufacturing cannot be covered adequately for individual products using LCA datasets, only material recycling, (thermal) recovery and disposal (landfill) are taken into account here.	–
Material masses in product ($mm_{p, EOL}$)	These parameters refer to the individual masses of materials directly contained in the product p. It is recommended to limit the considerations on materials representing major shares in terms of physical quantity, economic value or environmental impacts (as far as known).	kg

(continued)

---

[3]An increasing number of regulations force the original manufacturer to take back and recycle outdated products after use. As an example, Directive 2012/19/EU of the European Parliament on waste electrical and electronic equipment (WEEE) sets requirements for the treatment of electrical and electronic waste to be conducted by the producers (European Union 2012). As a consequence, the relevance of end of life processes is constantly growing for manufacturing companies.

Table 4.10 (continued)

Parameter	Description	Unit
Impact factors per material mass ($IF_{mm, EOL}$)	These parameters refer to impact factors characterizing recycling activities for contained materials in accordance with the selected end of life option. These are used to calculate environmental impacts for the treatment of all relevant material fractions by multiplication with the material masses mm contained in the product (compare Fig. 4.20). Additionally, datasets representing the production of potentially replaced primary materials can be used in order to award credits for the avoided resource demands (Baumann and Tillman 2004; Nicholson et al. 2009).[a]	e.g. kg CO_{2eq}/kg
Collection rate ($CR_{p, EOL}$)	The CR refers to the share of products that can be collected after the use stage and is thus available for end of life treatment (Baumann and Tillman 2004). Not collected products get lost somewhere in or after the use stage (Allwood 2012). They enter other local or global material routes or are even disposed in the natural environment. Consequently, the not collectable share of products is not further considered, even if in practice the products might be partly reused, recycled or disposed elsewhere.	%
Recycling rate ($RR_{p, EOL}$)	This parameter refers to incurring quantitative losses, if recycling is applied as end of life option.[b] Typical recycling rates for metals can for instance be found in Graedel et al. (2011), Gutowski et al. (2013a).	%

(continued)

Table 4.10 (continued)

Parameter	Description	Unit
Quality degradation rate ($QDR_{p,\,EOL}$)	This parameter refers to potential qualitative deterioration, i.e. a down-cycling of material. As an example, the purity of metals typically decreases with every recycling loop compared to primary metals (Allwood 2012). Rates can for example be derived from the ratio of market prices for primary and secondary material.[c]	%

[a]The allocation of end of life impacts allows for a high degree of methodological freedom. The described *substitution method* rewards credits due to the potential substitution of primary material by secondary material obtained from recycling. Thorough discussions of alternative principles are provided by (Nicholson et al. 2009; Allacker et al. 2014)

[b]Detailed discussions about related terms and approaches of (end of life) recycling as well as typical recycling rates for the main materials steel, aluminum, cement, paper and plastics can be found in (Gutowski et al. 2013a).

[c]According to the general allocation hierarchy provided in ISO 14044, allocations based on physical relations are the preferred option. If this is not possible, other relationships can be used such as the economic value (International Organization for Standardization 2006b). A detailed discussion about options to consider quality degradation in life cycle assessment can be found in (Kim et al. 1997).

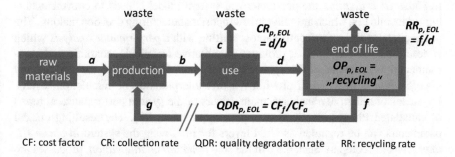

Fig. 4.19 Modeling of end of life stage, inspired by Baumann and Tillman (2004)

4.3.4 Contributions to Resource Efficient Manufacturing

The modeling phase with its inherent steps and methods provides decision support in building up system models of manufacturing systems from process up to value chain level. A progress beyond state of research can be stated in particular regarding the following aspects:

- *holistic modeling approach* from cradle to grave with a focus on production stage under consideration of all relevant resource flows from process level up to value chain level
- consistent, seamless and synergetic *coupling of MEFA, VSM, LCC and LCA* methods with underlying parameter list in terms of system modeling and performance evaluation to be used during data acquisition and model build up
- development of *load profile clustering (LPC)* methodology to automatically interpret dynamic resource demand profiles and quantify machine state-based resource demand levels

4.4 Phase II: Evaluation

In *phase II: evaluation* the parameterized system model is used to convert underlying data into PIs, which are suitable to support manufacturing decision making. The phase contains three consecutive steps, starting with a *performance analysis* which is described in Sect. 4.4.1. It provides quantitative technical, economic and environmental assessments of the system's initial state. In Sect. 4.4.2 the step *hotspot identification* is introduced, providing a first interpretation of calculation results. By means of sensitivity analyses, main drivers of the system performance in terms of calculated PIs are identified (compare requirement R7). The resulting critical parameters can be regarded as main levers for improving the system in *phase III: improvement* (compare Sect. 4.5). In the step *potential quantification,* processes are benchmarked with reference processes (compare Sect. 4.4.3), allowing to assess potential performance gaps (compare requirement R8).

4.4.1 Step 4: Performance Analysis

In *step 4: performance analysis* the initial state of the system under assessment is evaluated by means of the MEFA-based system model created in *phase I: modeling.* The performance analysis can be conducted on different manufacturing system levels, while an analysis on process level is regarded as starting point. Most indicators can then be aggregated on higher system levels by either addition or multiplication of lower level indicators. Their quantitative values highly depend on system level and reference flow (i.e. quantity, mass or surface size of products). As discussed in Sect. 2.1.3, appropriate PIs should be selected in accordance with the objectives of an analysis. Based on the methods that have been integrated into this concept, a basic system of performance indicators with focus on the production stage is presented in Fig. 4.20.

As depicted, the PI system builds upon a joint data basis and covers all three relevant evaluation dimensions: economic (compare Fig. 4.20a), technical (compare Fig. 4.20b) and environmental (compare Fig. 4.20c). Due to a focus on resource efficiency, nearly all PIs are related to physical resource inputs and outputs, i.e. on material, energy or labor. In accordance with the bottom-up modeling approach pursued, all PIs can basically be traced back to data collected on machine and process level. These time, quality or capacity related parameters do typically refer to a specific production resource (e.g. availability of a machine) or a combination of product and production resource (e.g. processing time of a certain product type on a specific machine). By aggregation and/or combination, resulting PIs on lower system levels like the energy intensity to perform a certain processing operation can often be transferred to higher system levels up to value chain level. PIs can – on all system levels – be allocated to different references such as products (e.g. energy intensity per product, per kg of product, per m^2 of product) or time periods (e.g. energy intensity per day, per month, per year). Economic PIs and environmental PIs can also be aggregated over the entire product life cycle. In order to collect the underlying data, supplementary information such as working times, production capacities or production schedules may sometimes be needed, which is not displayed here in order to maintain clarity.

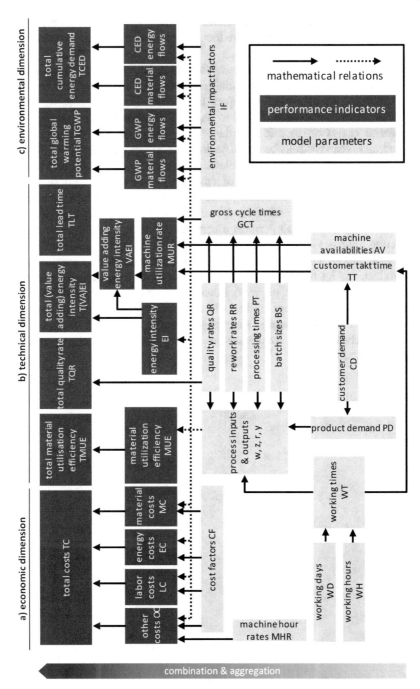

Fig. 4.20 Basic system of performance indicators for production stage considered in the concept

4.4.1.1   Technical Performance Indicator Calculation

In this concept, technical PIs basically build upon the traditional VSM methodology as introduced in Sect. 2.4.1, while extensions regarding resource efficiency are implemented. They cover information about direct and indirect energy demands as well as material utilization efficiency. Due to the adaptions and extensions applied, the developed method is referred to as *extended VSM (exVSM)* in the following. The technical PIs calculated and displayed for every process of the value stream are summarized in Table 4.11.

Table 4.11   Considered technical PIs

PI	description	unit
Processing time (*PT*) and lead time (*LT*)	The sum of all processing times (compare Sect. 4.3.3) on higher system levels is referred to as lead time. It is usually a highly relevant PI for producing companies (Bracht et al. 2011), revealing the total time that products need to cross the manufacturing system. Accordingly, it is typically linked to the overall material or product stock in the manufacturing system and hence also of economic relevance.	s/...
Net cycle time (*NCT*), gross cycle time (*GCT*)	The net cycle time *NCT* is a key performance indicator in traditional value stream mapping to express the capacity of a process (Erlach 2010). It can be calculated according to Eq. (4.7) $$NCT = \frac{PT_{proc}}{NOM_{proc} * BS_{proc}} \quad (4.7)$$ The gross cycle time *GCT* additionally takes into account the influence of defective parts and reworks due to insufficient process stability (compare quality rate *QR*), which reduce the throughput of affected processes. Further, actual availabilities of machines are considered, which are often reduced by maintenance or failure downtimes. Hence, the *GCT* allows for a more realistic assessment of existing production capacities (Erlach 2010).	s/...
customer takt time (*TT*)	The customer takt time *TT* indicates the required minimal frequency to manufacture a certain product in order to fulfil the customer demand (Erlach 2010). On factory level it can be calculated according to Eq. (4.8) $$TT = \frac{WD_f * WH_f}{CD_f} \quad (4.8).$$	s

(continued)

Table 4.11 (continued)

PI	description	unit
Machine utilization rate (*MUR*)	The machine utilization rate *MUR* is used as an easy to interpret indicator for the capacity use of production equipment. It gives indications about the ability of a process to satisfy the customer demand. Processes with the highest utilization rates along a process chain can be assumed to be bottlenecks, potentially limiting the overall production throughput (Banks et al. 2010). The *MUR* is calculated according to Eq. (4.9) $MUR = \frac{GCT}{TT}$ (4.9).	%
Quality rate (*QR*)	As described in Sect. 4.3.3, the quality rate *QR* is an input parameter of the system model. In the exVSM method, it is directly used as PI. Low *QR*s indicate the generation of defective parts, revealing resource wastage on process level. The later defective parts are produced along a process chain or value chain, the more significant are related effects due to induced resource demands in all previous manufacturing steps (Erlach 2010). By multiplying the quality rates of all processes *j*, a total quality rate *TQR* can be calculated according to Eq. (4.10) $TQR = \prod_{j=1}^{n} QR_j$ (4.10).	%
Material utilization efficiency (*MUE*)	In some processes, a certain share of raw material is intentionally or unintentionally removed from the product. Such losses are expressed through the material utilization efficiency *MUE*. It is calculated as the ratio of product output quantity y_j divided by raw material input quantity $w_{j,material}$ for each manufacturing process *j* (Yuan et al. 2012) according to Eq. (4.11) $MUE_j = \frac{y_j}{w_{j,material}}$ (4.11) In practice, removed material may be either treated as scrap, reused or recycled, which is not considered by the *MUE*. Achievable efficiencies highly depend on the chosen technology. In casting processes, for example, excess material to fill the gating systems results in *MUE*s of down to 30% (Allwood 2012; Heinemann 2016). Hence, both product and process redesign options can be taken into account to improve the *MUE*.	%
Energy intensity (*EI*)	In analogy to the EVSM method (compare Sect. 2.4.1), an energy intensity can be calculated for every process and/or product as well as on higher system levels. To increase the PI's usefulness, it can be calculated separately for all energy carriers, depending on the prevailing energy forms (e.g. electricity, heat). Generally expressed, the *EI* can be calculated as follows: $EI_j = \frac{w_{j,energy}}{y_j}$ (4.12) Energy demands related to TBS supply such as compressed air demands can be considered using energy intensity factors as indicated by Eqs. (4.5) and (4.6) in Sect. 4.3.3.2.	kWh/ …

(continued)

Table 4.11 (continued)

PI	description	unit
Value adding energy intensity (*VAEI*), non-value adding energy intensity (*NVAEI*)	The value adding energy intensity *VAEI* and non-value adding energy intensity *NVAEI* can be calculated for all processes with a sufficient data basis, i.e. if energy demands in different machine states are available (compare Table 4.7 in Sect. 4.3.3.2). In this concept, only energy demanded during non-productive machine times is declared as non-value adding, as this energy share is not related to any value adding activities (compare Fig. 4.7). In contrast, energy demanded by machine peripheries during processing such as pumps, filters, exhaust air system etc. is considered as value adding. *VAEI* and *NVAEI* can also be expressed as shares in relation to the total *EI*, while the sum of *VAEI* and *NVAEI* equals the total *EI* as defined before: $VAEI + NVAEI = EI$ (4.13) A low share of value adding energy can be regarded as indicator for existing saving potentials through optimized machine operation strategies (Herrmann and Thiede 2009; Teiwes et al. 2018)	kWh/ ...

4.4.1.2   Economic Performance Indicator Calculation

Beneath these technical indicators, economic PIs are derived applying life cycle costing (LCC) as introduced in Sect. 2.4.3. Based on resource flows mapped with the system model, a detailed cost analysis with cost allocation to different resource flows and processes can be conducted. Due to the case-specific relevance of resource flows, the following economic PIs are proposed as default, while additional cost related PIs can be defined as required. The explanations in Table 4.12 are focusing

Table 4.12 Considered economic PIs

PI	Description	Unit
Material costs (*MC*)	Material costs *MC* refer to costs for material inputs and outputs of processes and TBS. Potential negative cost flows, e.g. revenues due to sale of waste materials, are taken into account as well. The costs are averaged for all products manufactured over a certain period.	€/...
Energy costs (*EC*)	Energy costs *EC* are related to energy inputs of processes, also considering indirect demands of the TBS. For highly energy intensive processes, the shares of different energy forms contributing to the *EC* should be separately analyzed.	€/...
Labor costs (*LC*)	Labor costs *LC* refer to costs, which result from manual work. The focus is put on production personnel such as machine operators.	€/...

(continued)

Table 4.12 (continued)

PI	Description	Unit
Other costs (*OC*)	Other costs *OC* refer to further costs evolving from machine operation. With regard to a single machine, they can typically be received by multiplication of machine hour rate HR_{mach}, processing time PT_{proc} and product demand PD_{mach} $$OC = HR_{mach} * PT_{proc} * PD_{mach} \quad (4.14).$$	€/…
Total costs (*TC*)	Total costs *TC* refer to the total costs that are related to a specific product. For the production stage, the costs resulting from the inputs and outputs of materials, energies and labor as well as machine hour rates are equally allocated to all factory products of the same type, including expenses for defective parts and non-value adding energy demands $$TC = MC + EC + LC + OC \quad (4.15).$$	€/…

on the production stage, but similar calculations can also be conducted for the other life cycle stages.

4.4.1.3   Environmental Performance Indicator Calculation

Finally, environmental PIs are calculated, applying principles of the LCA methodology as introduced in Sect. 2.4.5. Therefore, the material and energy flows representing the life cycle inventory (LCI) first need to be translated into elementary flows. Subsequently, they can be transformed into environmental impacts in the step of life cycle impact assessment (LCIA), deriving flow specific midpoint LCIA results (Baumann and Tillman 2004). The method is typically applied by support of LCA software. In this concept, LCIA results in two mid-point impact categories (*GWP*, *CED*) for specific material and energy flows are taken from the *ecoinvent* database (compare Table 4.13). They are referred to as flow specific impact factors *IF* in the following. However, the consideration of additional environmental indicators is recommended, as *GWP* and *CED* do barely cover specific environmental issues which are not related to the generation and use of energy (e.g. acidification or toxicity). Further, GWP and CED are not independent from each other, which should be considered for PI weighting (compare Sect. 4.5.3).

Table 4.13 Selection of considered environmental PIs

PI	Description	Unit
Global warming potential (*GWP*)	The global warming potential *GWP* refers to the greenhouse gas emissions, which can be related to a specific product or process. The *GWP* has been selected due to its great global relevance (Rockström et al. 2009; Hauschild 2015) and related attention of politics and society as well as global climate agreements such as Kyoto protocol or Paris agreement (United Nations 1998, 2015). A calculation of the specific *GWP* for an input flow w_j is conducted as follows: $GWP_j = IF_{GWP} * w_j$ (4.16).	kg CO_{2eq}/...
Cumulative energy demand (*CED*)	The cumulative energy demand *CED* is an aggregated indicator, which basically refers to the total primary energy related with production, use and disposal of a good (Verein Deutscher Ingenieure 2015). In contrast to the direct energy demand during production stage *EI*, all energy but also material inputs are considered and converted to primary energy equivalents. The *CED* is regarded as a very useful indicator to supplement environmental impact assessment, even if a detailed impact assessment is performed (Klöpffer 1997). A calculation of the specific *CED* for an input flow w_j is conducted as follows: $CED_j = IF_{CED} * w_j$ (4.17).	kJ/...

4.4.1.4 Results visualization

On the one hand side, resulting PI values can be expressed mathematically in form of vectors. A *performance vector* $\vec{p}$, containing a set of PI values related to a system element or system level, shall be designated as follows:

$$\vec{p} = \begin{pmatrix} PI_1 \\ PI_2 \\ \dots \\ PI_i \end{pmatrix} \qquad (4.18)$$

In order to make results more tangible, a supplementary visual representation is proposed. In analogy to VSM analysis, value stream maps are used here for results visualization (compare Fig. 2.14 in Sect. 2.4.1). Such maps are regarded as a good basis for communication with other stakeholders and as starting point for system improvement (Erlach 2010). In order to identify synergies and drawbacks and to avoid problem shifting among PIs, all results coming from exVSM, LCC and LCA application are integrated into an *exVSM map* shown (compare Fig. 4.21). It provides a high degree of transparency regarding the initial performance of a factory and allows for a concurrent analysis of different PIs. An exVSM map for a single factory comprises detailed information on process level but also aggregated PIs representing process chain and factory level (compare value stream view in Fig. 4.21a). Total PIs can be calculated in relation to flexible reference values such as a factory product or a product mass. On process level, some indicators can be supplemented by graphical elements like amp displays, bar charts or pie charts in order to allow for a quicker

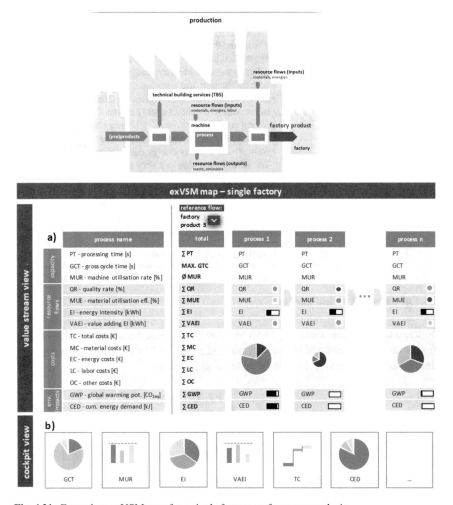

Fig. 4.21 Exemplary exVSM map for a single factory performance analysis

visual identification of resource wastage and potential hotspots. In addition, a *cockpit view* uses simple charts to show aggregated results (compare Fig. 4.21b).

The exVSM map provides a comprehensible overview from a local, factory internal perspective (gate to gate). In addition, a separate visualization is proposed to display results beyond the scope of single factories. In the value stream view (compare Fig. 4.22a), aggregated PIs for factories of the entire value chain and for other life cycle stages are displayed. While most technical PIs can only be calculated for the production stage, economic and environmental PIs can be provided for the entire product life cycle. The supplementary cockpit view (compare Fig. 4.22b) allows for a quicker visual analysis of individual contributions of each factory or life cycle stage to the overall results. It is particularly designed to support a first PI-based hotspot detection (compare Sect. 4.4.2.1).

In industrial practice, companies are – for comprehensible reasons – often not willing to share detailed information on process level as shown in Fig. 4.21. In

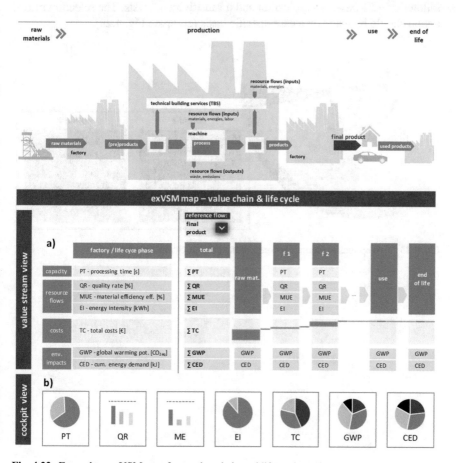

Fig. 4.22 Exemplary exVSM map for a value chain and life cycle performance analysis

contrast, aggregated results on factory level as presented in Fig. 4.22 are often regarded as less critical. Consequently, the acceptance of the proposed concept in industry can be increased, if aggregated results are shared with partners in a first step. Detailed results on process level can remain within the affected companies and be shared later to a necessary extend for system improvement.

4.4.2 Step 5: Hotspot Identification

The fifth step of the overall concept, *hotspot identification*, aims to identify main drivers regarding the overall system performance (addresses requirement R7). These hotspots are assumed to feature a certain potential for improvements in terms of resource efficiency. Two alternative options to identify hotspots are presented in the following: a PI-based identification and a sensitivity analysis. The selection of one option should be made on a case-specific basis (compare Fig. 4.23).

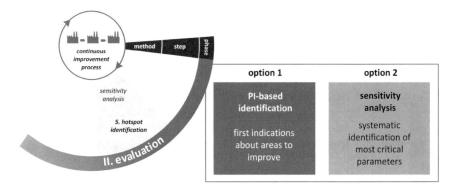

Fig. 4.23 Two options to identify hotspots along process chains or value chains

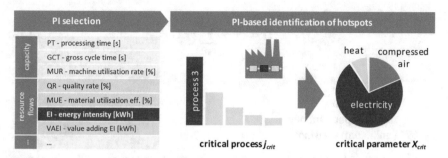

Fig. 4.24 Identification of hotspots by analyzing PIs exemplified for energy intensity

4.4.2.1 PI-Based Identification

The first option refers to a PI-based identification of rather obvious hotspots like the most relevant factories, processes or resource flows in terms of their contributions to certain PIs. This approach for hotspot identification is well-established and has been described by several other authors (Thiede 2012; Ghadimi et al. 2014; Heinemann 2016). It can be easily conducted based on *step 4: performance analysis* by a visual analysis of exVSM maps in order to derive first indications about areas to improve. As an example, processes can be compared regarding their contribution to the overall energy intensity *EI* on factory level as shown in Fig. 4.24. The comparison of energy demands on process level reveals that *process 3* is the dominating process in terms of energy utilization. This process can therefore be referred to as critical process j_{crit}. An analysis of the resource inputs identifies electricity inputs as main share contributing to this energy demand. Accordingly, the electricity input of *process 3* is referred to as critical parameter X_{crit}. Improvement measures to be implemented typically aim at improving critical parameters of critical processes. For the given example, a positive influence on other PIs such as total costs and environmental impacts can be expected as well.

4.4.2.2 Sensitivity Analysis

The identification of hotspots as described above does not necessarily help to find the right levers, i.e. the most effective spots to apply improvement measures. Therefore, a sensitivity analysis as second option to identify hotspots is proposed, which aims at a systematic identification of system model parameters with high influence on resulting PIs. A similar procedure in the context of resource efficient manufacturing is for instance proposed by Löfgren (2009). Basically, sensitivity analyses aim to quantify the sensitivity of a model and its respective outputs in relation to model input changes (Saltelli et al. 2008). Depending on model structure and complexity, system responds may be linear or non-linear, while stochastic effects are possible as well. A linear model behavior can for instance be expressed by Eq. (4.19).

$$Y_i = b_0 + \sum_{r=1}^{k} b_r X_r \qquad (4.19)$$

with:

Y_i model outputs
X_r model input parameters
b_0, b_r (unknown) constants

As shown in Fig. 4.25, sensitivity analyses typically follow a stepwise procedure: First, a design of experiments needs to be defined, i.e. relevant input parameters to change must be selected (compare Fig. 4.25a). Parameter values which shall be tested for each X_r then need to be defined (compare Fig. 4.25b), whereby the experiment is repeated in a defined interval. Input parameters should then be varied within the intervals and in accordance with realistic probability distributions as depicted in Fig. 4.25c (Saltelli et al. 2008). In the presented concept, this parameter variation follows a one-at-a-time (OAT) design, where only one parameter X_r of a set of parameters k is changed at a time. Potential effects of simultaneous multiple parameter changes cannot be identified by applying this method, but calculation efforts remain manageable. The effects of parameter changes to selected model outputs Y_i such as the total costs TC are then analyzed to draw conclusions about existing input–output-relationships (compare Fig. 4.25d). As results, for each examined Y_i an individual ranking of relevant input parameters can be obtained, while parameters with the highest impacts compared to their relative change are regarded as most critical. These critical parameters X_{crit} can be more or less any process and machine related model parameters described in Sect. 4.3.3 like process quality rates, rework rates, machine availabilities, resource input/output quantities etc. It is suggested to identify the parameters using sensitivity diagrams as indicated in Fig. 4.25. In analogy to the

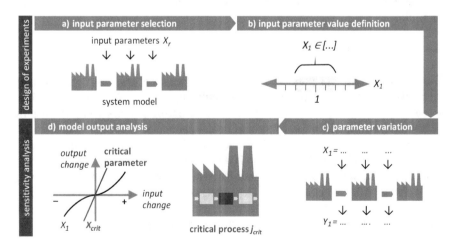

Fig. 4.25 Procedure for sensitivity analysis, inspired by Saltelli et al. (2008)

PI-based hotspot identification, processes featuring critical parameters are labeled as *critical processes*.

4.4.3   Step 6: Potential Quantification

Step number six of the presented concept aims at a *potential quantification* by means of benchmarking (addresses requirement R8). This method is well-established in industrial practice to detect improvement potentials. The procedure described in the following is oriented towards typical benchmarking procedures (e.g. Camp and Tweet 1994). Applications in the context of resource efficient manufacturing can for instance be found in Simon (2006), Tan et al. (2015), Mahamud et al. (2017) and the German standard DIN EN 16231, which provides a framework for energy efficiency benchmarking (Deutsches Institut für Normung 2012). In this concept, benchmarking addresses the critical parameters and critical processes as identified before. Through benchmarking, the performance can be put in relation to the performance of reference objects such as similar processes, resulting in performance gaps that represent realistically achievable improvement potentials (compare definition of resource efficiency in Sect. 2.2.2). Thus, the greater the assumed potential of a single process, the more priority should be given to improve it in subsequent steps. This principle is well applicable to the energy intensity *EI* of manufacturing processes as critical parameters due to good data availability in terms of reference values. However, a general transferability to other parameters is given, e.g. material inputs, quality rates or emissions. As displayed in Fig. 4.26, the potential quantification can be described as a four-step procedure in analogy to energy benchmarking.

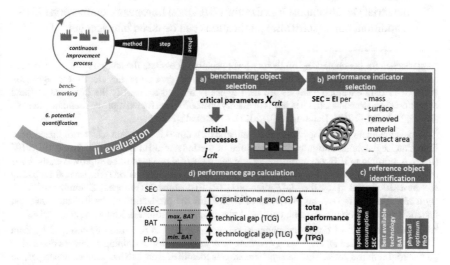

Fig. 4.26   Potential quantification by benchmarking *SEC* against *BAT* and *PhO*

(a) **Selection of benchmarking objects**: Although average data in terms of energy utilization on factory level is available for many branches (e.g. Lässig et al. 2016), factories are often regarded to be too complex and individual for a meaningful benchmarking with manageable efforts.[4] In contrast, single manufacturing processes are well-suited benchmarking objects due to a good availability of reference data. Thus, in this first step critical processes featuring energy as critical (input) parameter are selected as benchmarking objects.

(b) **Selection of PIs**: Performance indicators on process level should be used as metrics in order to identify inefficiencies and improvement potentials. However, PIs introduced in Sect. 4.4.1 such as the energy intensity are not directly usable for this purpose, as they do not contain detailed information about the manufacturing operations conducted. Hence, PIs need first to be related to a reference such as treated surface size or processed material mass. For an energy benchmarking, the *EI* can be put in relation to such a reference in order to obtain a specific energy consumption *SEC* (compare Sect. 2.1.3). If a value adding energy intensity *VAEI* has been calculated (compare Sect. 4.4.1), a value adding specific energy consumption *VASEC* can be derived as well.

(c) **Identification of reference objects**: A crucial step can be seen in the identification of reference objects to benchmark with. In this concept, best available technology *BAT*[5] and physical optimum *PhO* as two types of references are used in terms of energy efficiency benchmarking (Gutowski et al. 2013b). The *SEC* defined by *BAT* relates to technologies which are already in use and constitute best practice. *BAT* values are not absolutely fixed, but improve over time due to technological progress (compare technology factor in IPAT equation in Sect. 1.1). They are often formulated as ranges rather than as single values due to their dependence on product characteristics and processing parameters. Up to date reference values for *BAT* are for instance provided by United Nations Industrial Development Organization (2014) and European Commission (2018). In addition, other (scientific) publications can be taken into account.[6] The *SEC*

---

[4]Approaches for benchmarking on factory level usually necessitate the analysis of further system characteristics, e.g. climate conditions, production mix, product properties etc. Linear regression analyses or data envelope analyses are then typically applied to identify the influence of these characteristics. Demonstrations in terms of energy utilization can for example be found in Tan et al. (2015), Dehning (2017), Dehning et al. (2017), Flick et al. (2017).

[5]The best available technology (also best available technique) is a concept which originated from European Legislation. Currently, it is anchored in the Industrial Emission Directive 2010/75/EU (IED). According to IED, environmentally relevant industrial plants must be approved on the basis of *BAT*. The European Union publishes reference values regarded as *BAT* for several industrial sectors and activities in so called *BAT* reference (BREF) documents (European Commission 2018). By using *BAT*, energy saving potentials of 9–30% for the production of the five most relevant materials steel, aluminum, cement, paper and plastics can be estimated (Gutowski et al. 2013a).

[6]A detailed analysis of energy benchmarks on machine level is for instance presented by Zein (2013), who identified *efficient energy consumption functions* for machining processes depending on material removal rates. Thus, an *efficient edge* is identified from production theory perspective (Dyckhoff 2006). As a result, a distinct *BAT* value can be provided for the specific process parameters required to fulfill the process target (e.g. reach a certain surface roughness). Overviews for typical

defined by the *PhO* relates to the minimal energy input which is necessary to conduct a process from a physical perspective (Volta 2014; Keichel et al. 2015; Kreitlein et al. 2017). It can be calculated based on physical relations and therefore represents the absolute possible minimum energy input. The *PhO* can rather serve as general orientation than as target value, as it is not realistically achievable in practice. However, improvements towards *PhO* can for instance be made by research breakthroughs and by implementing cutting-edge technologies (Gutowski et al. 2013b). As indicated in Table 4.14, reference values can be found for diverse types of manufacturing processes. Due to the diversity of applied processing parameters and processed materials, the given range for *BAT* is quite broad for some processes like turning and milling (Duflou et al. 2012). Therefore, applied processing parameters (e.g. cutting speed, feed rate, material properties) should be taken into account in order to improve the validity of the comparison.

(d) **Calculation of performance gaps**: The absolute deviations between reference values provided by *PhO* and *BAT* with the actual measured performance *(SEC, VASEC)* of the selected processes are calculated. Such calculations allow to distinguish between different kinds of saving potentials. Hence, the resulting performance gaps *PG* can be expressed using different indicators. The *organizational gap (OG)*, as the performance gap between measured *SEC* and *VASEC* for a process *j*, can be interpreted as achievable saving potential due to organizational reasons. It can be traced back to non-value adding machine operation times (compare definition of non-value adding machine states in Sect. 4.3.3):

$$OG_j = SEC_j - VASEC_j \qquad (4.20)$$

A *technical gap (TCG)* between *VASEC* and *BAT* according to Eq. (4.21) can be regarded as additional achievable saving potential due to (mainly) technical improvements by using available solutions. *BAT* values should be selected in accordance with applied processing parameters, which exert significant influence on achievable *SEC* (Zein 2013). If *BAT* values are given as range, a distinction between a *minimal technical gap* and a *maximal technical gap* can be conducted to reflect uncertainties.

$$TCG_j = VASEC_j - BAT_j \qquad (4.21)$$

A *technological gap (TLG)* between *BAT* and *PhO* is suitable to make statements about the general efficiency of the chosen technology. For processes featuring great technological gaps, alternative process technologies should be taken into account. Apart from that, improvements beyond *BAT* can also be achieved by creating new combinations of best practice solutions or by developing entirely new, innovative approaches (Simon 2006).

SECs of various manufacturing processes depending on the process rate – but not necessarily reflecting *BAT* – are presented by Gutowski et al. (2006, 2013b).

Table 4.14 Exemplary *BAT* and *PhO* references in terms of energy demands for selected manufacturing processes

Process type	*BAT*[a]	*PhO*[b]
Injection molding	0.1–5.8 kWh/kg	$\Delta H_m(T_2) - \Delta H_m(T_1)$ with: $\Delta H_m(T_1)$ melting enthalpy at initial temperature $\Delta H_m(T_2)$ melting enthalpy at target temperature.
Die bending	0.02–0.06 kWh/100t	$\frac{K_{form}*\sigma_m*l*t_{wp}^2}{w} * d$ with: K_{form} constant for die opening distance σ_m ultimate tensile strength l length of bend t_{wp}^2 stock thickness d die closed depth.
Turning, milling	0.0001–0.05 kWh/cm^3	$\frac{k_{c1.1}}{a_p^{m_c}} * K_V * K_\gamma * K_{TM} * K_{TW} * K_L * K_{WF} * V_{chip}$ with: $k_{c1.1}$ specific cutting force m_c chip thickness exponent a_p chip thickness K_V correction factor for cutting speed K_γ correction factor for cutting angle K_{TM} correction factor for tool material K_{TW} correction factor for tool wear K_L correction factor for workpiece shape V_{chip} removed chip volume.
Hard chrome plating	0.95–3.56 kWh/m^2 * μm	$\Delta H_f^{chrome} * \frac{m_{chrome}}{M_{chrome}}$ with: ΔH_f^{chrome} net enthalpy of chrome formation m_{chrome} mass of deposited chrome M_{chrome} molar mass of chrome.
Hardening	n.a	$m * c_p * (T_2 - T_1)$ with: m mass of workpiece c_p specific heat capacity of material T_1 initial temperature T_2 target temperature.

[a]Ref.: Bayerisches Landesamt für Umweltschutz (2003), Gutowski et al. (2006, 2007), Thiriez and Gutowski (2006), Duflou et al. (2011, 2012), Yoon et al. (2014), Ecoinvent Centre – Swiss Centre for Life Cycle Inventories (2018)
[b]Ref.: Kalla et al. (2009), Böge and Böge (2015), Millot et al. (2015), Kreitlein et al. (2017)

$$TLG_j = BAT_j - PhO_j \tag{4.22}$$

The *total performance gap (TPG)* between measured *SEC* and *PhO* represents the maximum saving potential which could be achieved, as long as the performed manufacturing task does not change. Further (theoretical) saving potential could only be achieved by a product re-design.

$$TPG_j = SEC_j - PhO_j \tag{4.23}$$

All performance gaps can be expressed in both absolute numbers but also relative to the *SEC*. On the one hand, absolute gaps help to quantify technical, economic and environmental improvement potentials. Relative performance can, on the other hand, give indications about the probability to achieve improvements.

The described procedure for potential quantification is summarized in Fig. 4.26. Similar procedures to quantify saving potentials can for instance be found in Rahimifard et al. (2010) as well as in Drumm et al. (2013), who describe a stepwise energy loss cascade between *SEC* and *PhO*.

4.4.4   Contributions to Resource Efficient Manufacturing

The evaluation phase with its inherent steps and methods provides a progress beyond state of the art by combining different complementary methods for performance assessment and evaluation. The following aspects are regarded as important contributions towards improved decision support:

- provision of a *performance indicator system* for a multi-criteria assessment with regard to technical, economic and environmental aspects of manufacturing
- development of integrated *extended VSM maps* for gate to gate as well as cradle to grave scopes, allowing for an improved system understanding through a clear and comprehensible visualization of results
- application of *sensitivity analyses* to prioritize fields of action by identifying critical processes and critical parameters, which are regarded as main levers to apply improvement measures
- development of a *benchmarking method* to identify and quantify resource saving potentials on process level by calculating performance gaps against best available technology and physical optimum

4.5   Phase III: Improvement

In *phase III: improvement*, the focus is put on the identification, evaluation and selection of measures to improve the system performance. Starting with Sect. 4.5.1, the challenge to find suitable measures is addressed by the application of knowledge-based systems (KBS), reducing the need for expert involvement (addresses requirement R9). The subsequent step *measure evaluation* as described in Sect. 4.5.2 makes use of the existing system model to conduct parameter studies. As a result, impacts of specific measures on the system performance can be forecasted (addresses requirement R10). Finally, in Sect. 4.5.3 a procedure for *measure selection* is presented to select best fitting measures in accordance with individual objectives (addresses requirement R11). In order to solve potential target conflicts, MCDA methods are used for final decision making (compare Sect. 2.3.3).

4.5.1 Step 7: Measure Identification

Step seven of the overall concept, *measure identification*, deals with strategies to find specific improvement measures. Those shall in particular address critical processes and critical parameters featuring significant improvement potentials according to *step 6: potential quantification*. As described in Sect. 2.2.1, common strategies of measure identification build on rather general improvement principles, which need to be transferred to a specific use case, e.g. by expert consultation. However, the transferability of measures from one system to another must be assured. In order to facilitate measure identification, a measure database as part of a knowledge-based system (KBS) can be used (Blume et al. 2018). Such KBS as a type of decision support systems are designed to support human actors in decision situations by providing recommendations based on documented expert knowledge (compare Sect. 2.3.4). As displayed in Fig. 4.27, they typically contain three elements (Puppe 1991; Kwon et al. 2005):

- a *knowledge database* contains declarative knowledge (e.g. established solutions for specific problems) as well as procedural knowledge, i.e. rules to apply the declarative knowledge
- an *inference machine* applies rules from the knowledge database to the decision situation

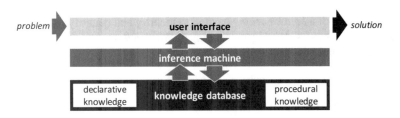

Fig. 4.27 Main elements of a knowledge-based system, own illustration based on Puppe (1991), Kwon et al. (2005)

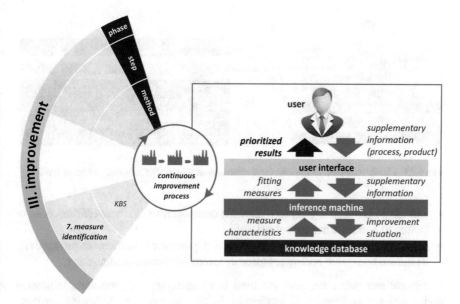

Fig. 4.28 Overview of KBS to identify improvement measures

- a *user interface* allows to feed the KBS with data and is capable to display obtained results

The KBS developed in this concept is structured accordingly (compare Fig. 4.28). A user, e.g. a production manager, can interact with the system through a user interface in order to identify suitable improvement measures. The inference machine is capable to match the current improvement situation with measures that are saved in the knowledge database. Results are then communicated to the user. Depending on the situation, the user may need to provide supplementary information about the process or product in order to improve the quality of results.

4.5.1.1 Measure Description in Knowledge Database

The knowledge database contains a set M of alternative improvement measures $m_i \in M$ for manufacturing systems with a main focus on process level. All measures are described in a systematic way. To create the database, case specific and domain specific measures are taken from literature and from expert consultation (Blume et al. 2018). Based on the work of Despeisse (2013) and Zhou et al. (2013), the improvement measures are structured according to four major underlying improvement strategies:

- *change resource,* e.g. substitute a resource for an eco-friendlier one
- *manage resource,* e.g. synchronize waste generation and resource demands to allow a reuse of resources

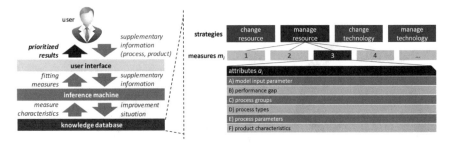

Fig. 4.29 Characterization of improvement measures in knowledge database as part of KBS

- *change technology*, e.g. replace machine components with more efficient ones or switch to a more efficient technology
- *manage technology*, e.g. change processing parameters to increase productivity or reduce non-value adding resource demands by automatic shut-down

Several *attribute categories* are used to characterize each measure, indicating under which conditions it is applicable. In the database, the following attribute categories with non-numerical values are typically used (compare Fig. 4.29):

(a) A *model input parameter* primarily tackled by the measure: In analogy to *step 5: hotspot identification* this can relate to a resource, process or machine related parameter like energy input quantity, material input quantity, processing time, machine availability etc.

(b) A *performance gap (PG)*, which is mainly addressed by the measure: In analogy to *step 6: potential quantification* this relates to either an organizational, technical or technological *PG*.

(c) A measure's general applicability regarding different *process groups*: A certain degree of transferability is assumed within the same process group of manufacturing according to DIN 8580 (compare Sect. 2.1.2).

(d) A measure's applicability in terms of *process types*: The transferability between processes from the same process group of manufacturing according to DIN 8580 is assumed to be even higher. As an example, an energy saving mode may be generally suited for machines conducting all process groups and types, while an efficient spindle drive is only applicable to machine tools featuring a spindle, i.e. to certain processes from the main group *separating* such as milling, turning or grinding.

(e) A measure's requirements in terms of other *process parameters*: Further requirements to apply a measure may address parameters such as process temperature, process speed etc. Additional information may be needed to check the fulfilment of these attributes, which can be demanded through the user interface.

(f) Further requirements in terms of *product characteristics*: For the application of some measures, characteristics of the processed product such as main material or product dimensions might be highly relevant.

Table 4.15 Excerpt of the knowledge database containing different improvement measures

Strategy	Measure	Model input parameter	Performance gap	Process category	Process type	Process parameters and product characteristics
Change resource	Use alternative non-mineral oil based cutting fluids	Material input: coolant and lubricant	TLG	Separating	All	Coolant and lubricant input > 0
Manage resource	Collect scrap material for recycling	Material inputs	TCG	All	All	Material waste output > 0
Change technology	Inductive hardening instead of furnace hardening	Energy inputs	TLG	Change material properties	Thermal treatment	–
Manage technology	Switch from continuous operation to batch operation	Energy inputs	OG	All	All	Batch size BS = 1 VAEI/EI < 1

An excerpt of the knowledge database is presented in Table 4.15, providing measure examples that address different improvement strategies.

In order to allow for an understanding of the logics behind the automatic measure identification, a mathematical measure description is provided. Characteristics of each measure are expressed as vector $\vec{a}_m$:

$$\vec{a}_m = \begin{pmatrix} a_1 \\ a_2 \\ a_3 \\ a_4 \\ a_5 \\ a_6 \end{pmatrix} = \begin{pmatrix} model\ input\ parameter \\ performance\ gap \\ process\ group \\ process\ type \\ process\ parameter \\ product\ characteristic \end{pmatrix} \quad (4.24)$$

To enable a filtering in accordance with a measure's suitability (compare Sect. 4.5.1.2), numerical values for each attribute category are required. Thus, a further breakdown is useful at this stage. As an example, the attribute category a_2 *performance gap* can contain three sub-attributes a_{2_1} *organizational gap (OG)*, a_{2_2} *technical gap (TCG)* and a_{2_3} *technological gap (TLG)*. Thus, a general attribute vector $\vec{a}_m$ with numerical attribute values can be redefined as follows:

$$\vec{a}_m = \begin{pmatrix} a_{1\_1} \\ a_{1\_2} \\ \cdots \\ a_{i\_j} \end{pmatrix}; \quad a_{c\_d} = \{0, 1\} \quad c, d \in N \tag{4.25}$$

with:

a measure attribute
c index of measure attribute category
d index of measure sub-attribute

Each sub-attribute can then be assigned a numerical value, i.e. either 1 for criteria fulfilment or 0 for non-fulfilment.

Example

A measure aims to reduce the energy input of a process by replacing a component of a separating machine with a more energy efficient one. It can be described with the following attribute vector, whereby all not displayed attributes are assumed to be 0:

$$\vec{a}_1 = \begin{pmatrix} a_{1\_1} \\ a_{1\_2} \\ \cdots \\ a_{2\_1} \\ a_{2\_2} \\ a_{2\_3} \\ \cdots \\ a_{3\_2} \\ a_{3\_3} \\ a_{3\_4} \\ \cdots \end{pmatrix} = \begin{pmatrix} 1 \\ 0 \\ \cdots \\ 0 \\ 1 \\ 0 \\ \cdots \\ 0 \\ 1 \\ 0 \\ \cdots \end{pmatrix} \quad \text{with:} \quad \begin{aligned} & a_{1\_1} \; energy\; input\; quantity \\ & a_{1\_2} \; material\; input\; quantity \\ & a_{2\_1} \; organizational\; performance\; gap \\ & a_{2\_2} \; technical\; performance\; gap \\ & a_{2\_3} \; techno\log ical\; performance\; gap \\ & a_{3\_2} \; reshaping \\ & a_{3\_3} \; separating \\ & a_{3\_4} \; joining \end{aligned}$$

$$\tag{4.26}$$

4.5.1.2   Measure Retrieval by Inference Machine

The inference machine is intended to identify the most suitable measures based on the described measure characteristics. Therefore, it applies procedural knowledge in conditional form, comparing measure attributes and improvement situation. A description of the improvement situation is taken from the system model after completion of *phase II: performance evaluation*. It is characterized by similar attributes as improvement measures. As an example, the model input parameter here refers to a critical parameter identified in *step 5: hotspot identification* that opens the possibility to improve a certain PI (e.g. s_{1_1} for energy input as critical parameter to reduce the total costs). As further example, the attribute s_{2_1} now refers to an organizational gap, caused by a high share of non-value adding energy, which needs to be tackled. Accordingly, the improvement situation can be described as vector $\vec{s}$ in analogy to the attribute vector $\vec{a}_m$ of each measure:

$$\vec{s} = \begin{pmatrix} s_{1\_1} \\ s_{1\_2} \\ \cdots \\ s_{i\_j} \end{pmatrix}; \quad s_{c\_d} = \{0, 1\} \text{ and } c, d \in N \quad (4.27)$$

In order to identify measures that are suited for the current improvement situation, a comparison of vectors $\vec{s}$ and $\vec{a}_m$ is performed for every measure contained in the knowledge database. Therefore, a vector multiplication is conducted in a first step to calculate the inner vector product, returning a value of *1* or *0* for each summand $s_{c\_d} \cdot a_{c\_d}$:

$$\vec{s} \cdot \vec{a}_m = (s_{1\_1} \cdot a_{1\_1} + s_{1\_2} \cdot a_{1\_2} + \cdots + s_{c\_d} \cdot a_{c\_d}) \quad (4.28)$$

To calculate an overall level of matching between measure and improvement situation, the number of relevant attributes must be defined. In some cases, the improvement situation demands for the fulfilment of exactly one specific sub-attribute from an attribute category. This applies for sub-attributes from the categories *model input parameter*, *performance gap*, *process category* and *process type*. They shall be denominated as set A_s. Only one match per category is aspired, i.e. $\left| A_s \right| = 4$. As an example, the improvement situation may demand for a measure that tackles energy input as model input parameter. It is neither regarded as benefit nor as drawback, if a measure addresses other input parameters. In other cases, the attribute category may contain several attributes of a measure that should be fulfilled by the improvement situation. This applies for the categories *process parameter* and *product characteristic*. The number of additional attributes to be fulfilled is individual for each measure m and shall be denominated as set A_m. Here, a total matching of attributes between measure and improvement situation is rewarded. As result, a relative matching level *ML* can be calculated through division of the inner product of vectors $\vec{s}$ and $\vec{a}_m$ by the total number of relevant attributes to consider for the matching:

$$ML = \frac{\vec{s} \cdot \vec{a}_m}{A_s + A_m} \quad \text{with } ML \in [0; 1] \quad (4.29)$$

The inference machine calculates a matching level for each measure from the knowledge database for the existing improvement situation, resulting in a list of measures sorted by their *ML*. A measure that receives a maximal *ML* of *1*, i.e. all attributes are matching between improvement situation and measure, is referred to as

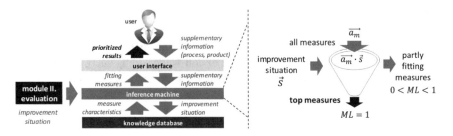

Fig. 4.30 Measure retrieval by inference machine for a given improvement situation

top measure. From the perspective of statistics, this matching resembles the calculation of a *Jaccard index* for binary attributes (Eckey et al. 2002). It is used to gauge the
similarity between finite sample sets, in this case between the vectors representing
improvement situation and measure characteristics. Compared to other metrics for
similarity analyses such as the *simple matching coefficient,* it only considers characteristics that are mutually fulfilled (present) and neglects characteristics that are
mutually not fulfilled (absent). The procedure of measure retrieval by the inference
machine is summarized in Fig. 4.30.

4.5.1.3   Measure Visualization by User Interface

The results are displayed in a user interface with reference to related PIs, critical
parameters and critical processes that describe the improvement situation (compare
Fig. 4.31). The list is, on the one hand, sorted by the impact of critical parameters
on each corresponding PI. Additionally, the measures are sorted according to the
absolute size of the addressed performance gap in order to provide an advanced
prioritization. However, these sortings cannot reliably forecast a measure's impact
on the corresponding PI. For this reasons, further calculations are conducted in *step
8: measure evaluation.*

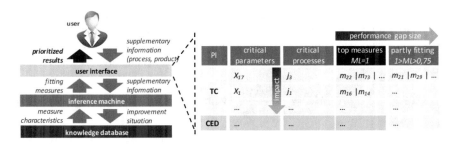

Fig. 4.31 Visualization of measures identified by the KBS for a given improvement situation

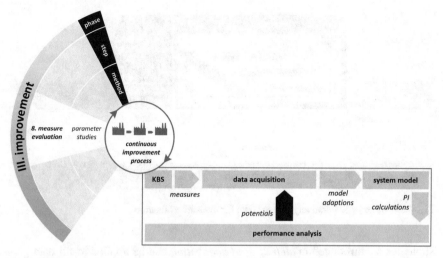

Fig. 4.32 Procedure for measure evaluation

4.5.2 Step 8: Measure Evaluation

In the step of *measure evaluation,* suitable measures identified by the KBS are evaluated with regard to their expected improvement potentials. Therefore, one or multiple system parameters as described in *step 3: system modeling* are typically changed in the system model (compare Fig. 4.32). Further, changes in the system structure might be applied as well, for instance if several process steps are integrated into one new process. Corresponding changes in system performances on all system levels can then be compared with the initial state. As measures are at first assessed independently from each other, potential interrelations and synergies between different measures may remain undetected. Therefore, it is recommended to evaluate promising measures simultaneously in a second step, at latest before measure implementation. In addition, cross-impact analyses[7] could be applied in order to systematically assess these interrelations. Otherwise, expected improvements may not fully materialize in practice (Hesselbach 2012).

In order to support users in the change of system parameters, the KBS contains original references for many measures, e.g. web links or literature references. In some cases, these can already provide information about parameter changes to be applied to the system model, e.g. reductions of energy input on process level, improvements of quality rates or reductions of scrap materials. Still, the transferability of quantitative values should always be critically questioned due to potential differences in the underlying system characteristics. This can be a difference in machine dimensioning, different process parameters etc. Apart from the original references, data acquisition

---

[7]The method of cross-impact analysis goes back to the 1960s (Gordon 1994). It was mainly developed to predict the probability of future events and their correlations. An example in the context of resource efficient manufacturing can be found in Dehning (2017).

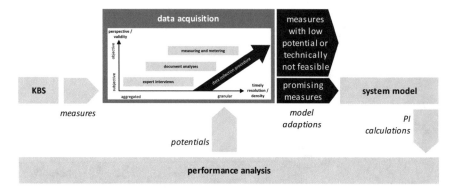

Fig. 4.33 Strategy to iteratively acquire data for measure evaluation

strategies as introduced in *step 2: data acquisition* can be applied to fill data gaps (compare Fig. 4.6 in Sect. 4.3.2):

- *Expert consultations* might be useful to quickly estimate the influence of measures to certain system parameters. In particular for measures addressing the technical performance gap *TCG* (compare Sect. 4.4.3), experts can often provide reasonable first estimations based on their experience knowledge.
- *Document analyses* can help to further validate expert appraisals. Machine documentations can for instance be used to judge about the energetic relevance of single components for the overall resource demand of a machine (Zein 2013).
- Additional *meterings and measurements* might be required for some system parameter changes due to a lack of experience or documented knowledge. As an example, the assessment of a heat recovery system usually requires information about inlet and outlet temperatures for both hot and cold media as well as related flow rates (Simeone et al. 2016). If such data is not directly available, corresponding measurements should be conducted during regular machine operation. Further, some measures may also necessitate additional experiments, e.g. measures suggesting adapted process parameters in order to shorten processing times or to increase product quality. Experts should be involved in the measurements in order to acquire all relevant measurands.

Universal guidelines to acquire data for measure evaluation can hardly be derived, as the need for data is highly individual for each improvement measure. In general, an iterative procedure starting with roughly estimated data is recommended as indicated in Fig. 4.33. Efforts for data acquisition should be kept reasonable, i.e. data quality for promising measures should be gradually increased. For measures featuring high expected improvements, data should be validated thoroughly by additional measurements or consultation of further internal or external experts. Obtained results can then be taken as input for a final ranking and the selection of improvement measures (compare Sect. 4.5.3).

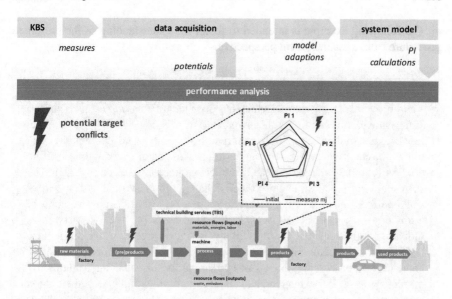

Fig. 4.34 Performance analysis and potential resulting target conflicts

A change of the system structure or parameters usually entails changes in the system performance, which can be assessed by means of the developed PI system (compare Sect. 4.4.2). Changes can be either positive or negative for each PI, potentially resulting in target conflicts. On the one hand, measures may lead to unbalanced improvements and deteriorations of different target dimensions or PIs respectively. On the other hand, target conflicts can also occur regarding benefits and drawbacks for different factories of a value chain. Additionally, also a problem shifting between different life cycle stages may occur. These different kinds of target conflicts involving multiple manufacturing system levels are indicated in Fig. 4.34. A visual analysis, e.g. using spider web charts, allows for a first comparison of different measures with the initial situation or against each other. In order to solve target conflicts and take the best decision in line with aspired targets, multi-criteria methods can be applied as described in *step 9: measure selection.*

The following example is intended to foster the understanding of the presented procedure from a mathematical perspective.

Example

An improvement measure m_1, which has been identified as *top measure* in step 7: measure identification, shall be evaluated from a single factory perspective. In the course of data acquisition, expert consultations reveal a potential reduction of the electrical power demand during processing of ~72 kW for a specific process (compare machine related model parameters in Sect. 4.3.3.2). The corresponding input parameter value is changed in the virtual system model and a performance recalculation is carried out. As result, PI improvements in terms of energy intensity EI (-20 kWh), total costs TC (-2.5 €) as well as global warming potential GWP (-10 kg CO_{2eq}) on process as well as factory level are expected. However, at the same time the measure is supposed to imply an increase in processing time PT of 100 s on process level and in lead time LT on factory level. Performance changes between initial and changed situation can now be compared using performance vectors as introduced in Sect. 4.4.1. Comparing the initial performance vector $\vec{p}_{init}$ with the new performance vector $\vec{p}_{m_1}$ by subtraction underlines the existing target conflict on factory level:

$$\vec{p}_{change} = \vec{p}_{init} - \vec{p}_{m_1} = \begin{pmatrix} PT_{init} - PT_{m_1} \\ QR_{init} - QR_{m_1} \\ MUE_{init} - MUE_{m_1} \\ EI_{init} - EI_{m_1} \\ PC_{init} - PC_{m_1} \\ GWP_{init} - GWP_{m_1} \\ CED_{init} - CED_{m_1} \end{pmatrix} = \begin{pmatrix} -100 \\ 0 \\ 0 \\ 20 \\ 2.5 \\ 10 \\ 0 \end{pmatrix} \quad (4.30)$$

By normalization of $\vec{p}_{m_1}$ in relation to $\vec{p}_{init}$, relative changes can be made visible in order to allow for a better comparison of different PIs. A visualization by a spider chart can foster the comprehensibility of results (compare Fig. 4.35). For the PIs quality rate and material utilization efficiency reciprocal values are used in the chart in order to achieve a uniform direction of optimization.

Apparently, the measure entails the highest relative improvements for the PI energy intensity.

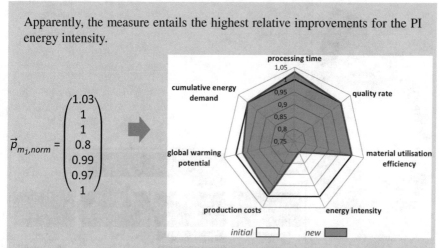

$$\vec{p}_{m_1,norm} = \begin{pmatrix} 1.03 \\ 1 \\ 1 \\ 0.8 \\ 0.99 \\ 0.97 \\ 1 \end{pmatrix}$$

Fig. 4.35 Evaluation of exemplary measure against initial state and resulting target conflict

4.5.3   Step 9: Measure Selection

The final step *measure selection* describes a procedure to select the *best* measures for implementation from perspective of a decision maker, considering individual preferences regarding evaluation dimensions and PIs. The set of PIs defined for an analysis may contain several indicators, all featuring a certain relevance for the evaluation of current and future system performance. Though, not all PIs might have the same importance for every manufacturing decision. Therefore, a weighting of PIs, i.e. an assessment of their importance expressed through a weighting factor, is regarded as a prerequisite for measure selection (compare Fig. 4.36a). As an example, the total costs per part might be the main indicator for decision making in a specific situation. Other PIs may also be used for hotspot identification and measure identification, whereas they play a minor role for deciding about the implementation of a measure. In addition, potential overlaps between PIs should be taken into account for weighting in order to avoid biased decisions. For instance, the PI energy costs is a subset of total costs (compare Sect. 4.4.1). If both are weighted equally with all other PIs, energy costs have an unintentional large influence on the decision. After weighting PIs according to their relevance, one or several measures from a set of possible measures must be selected. Against the background of multiple target dimensions and PIs considered, this problem requires techniques of multi-criteria decision analysis (MCDA) as presented in Sect. 2.3.3. Two alternative options are proposed to further process PI values within this concept (compare Fig. 4.36b): a measure comparison based on weighted sums or based on the PROMETHEE method.

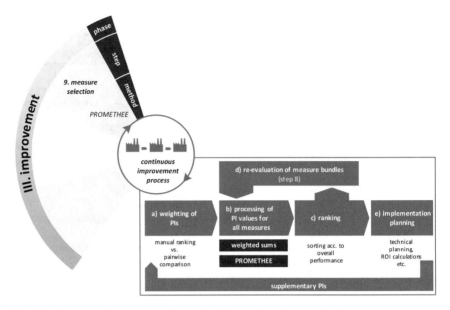

Fig. 4.36 Procedure for measure selection including an iterative re-evaluation

The decision for one of the two alternative procedures should be taken individually for each use case, depending on the complexity of the decision situation and the clarity of results in *step 8: measure evaluation*. As result, a ranking of measures from best to worst is received in accordance with their overall use values (compare Fig. 4.36c). In order to gain transparency about potential interdependencies of measures, a re-evaluation of promising measure bundles is recommended by iteratively repeating *steps 8* and *9* (compare Fig. 4.36d). The best measures or measure bundles according to the received ranking should then be selected for a further implementation planning (compare Fig. 4.36e). This includes a thorough technical feasibility study followed by a detailed technical planning on the one hand and additional investment calculations[8] on the other hand, e.g. considering return on investment (ROI) and payback calculation. Supplementary PIs evolving from implementation planning can also be fed back into multi-criteria evaluation. However, related aspects are not elaborated here, as a meaningful procedure is highly individual for each use case and the actual implementation of measures is not in the focus of this concept.

---

[8]The discipline of investment calculations, also referred to as capital budgeting, applies static and dynamic methods in order to judge about the financial consequences of invests. Advanced methods do also consider risks of future developments. A broad overview about common methods can be found in Kruschwitz (2014).

4.5.3.1 Measure Ranking By Weighted Sums

A measure ranking using weighted sums is proposed for a quick analysis and for rather simple cases, i.e. a manageable number of alternatives, few or no target conflicts etc. Corresponding examples regarding resource efficient manufacturing can for instance be found in Thiede (2012) and Dehning (2017). The procedure starts with a manual weighting of PIs. Therefore, every PI i is assigned a weighting factor w_i, whereby the sum of all weights equals 1. In analogy to the performance vector, weights can be expressed as vector $\vec{w}$:

$$\vec{w} = \begin{pmatrix} w_1 \\ w_2 \\ \dots \\ w_i \end{pmatrix} \quad \text{where} \quad \sum w_i = 1 \qquad (4.31)$$

with:

w_i individual weight of PI_i
I total set of considered PIs

PI weights can be either chosen equally for all indicators or individual for each PI in order to set priorities (e.g. total costs are more relevant than total processing time). The PI weighting bases upon the individual judgement of the decision maker. Subsequently, a normalization of PI values is conducted with respect to the initial state to achieve a better comparability of indicators with different scales. Finally, an aggregation of normalized values for each alternative measure to one single indicator under consideration of PI weights is carried out to receive a performance score PS. Mathematically, this is achieved by multiplication of normalized performance vector and weighting vector:

$$PS = \vec{p}_{m_i,norm} * \vec{w} = (PI_1 w_1 + PI_2 w_2 + \cdots + PI_i w_i) \qquad (4.32)$$

Based on the numeric PS values, a ranking from best to worst can be realized (Triantaphyllou 2000). It can be used to set priorities for measure implementation.

4.5.3.2 Measure Ranking by PROMETHEE Method

As an alternative, an application of advanced MCDA methods allows for a more systematic ranking of alternatives. This option is suitable for rather complex decisions with many improvement options and occurring target conflicts. Further, this option is recommended before finally implementing measures in the real-world system. Examples in the context of resource efficiency and sustainable manufacturing can for instance be found in Ilgin and Gupta (2010), Ude (2010), Walther (2010) and Haag (2013). Multi-attribute decision making (MADM) methods are required here, as a selection from a distinct set of solutions is made (compare Sect. 2.3.3). PROMETHEE has been identified as a suitable method to be applied in this concept. Compared with other advanced MADM methods, it is rather simple to use and has therefore reached a high popularity among practitioners (Behzadian et al. 2010). A detailed discussion about benefits and drawbacks of PROMETHEE as well as similar MADM methods can be found in Reinhardt (2013). Figure 4.37 shows the procedure to be followed.

The procedure starts with a weighting of PIs. Weighting factors are manually assigned to each PI (compare Fig. 4.37a). Subsequently, pairwise comparisons between all identified alternative measures need to be conducted for each PI, i.e. the absolute PI values for each alternative measure are subtracted from each other (compare Fig. 4.37b). As a result, deviations between all alternatives for each PI are received, e.g. the deviation in costs related with two alternatives. Hence, the consequences of each alternative must be known and quantified. In step three, for every attribute a *preference function* is defined (compare Fig. 4.37c). It reflects the decision maker's preference of one alternative towards another, depending on the

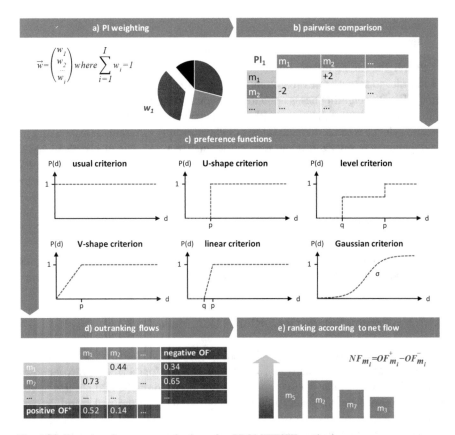

Fig. 4.37 Procedure for measure selection using PROMETHEE method

absolute difference in a criterion's value. Accordingly, one of the following preference relations must be assigned to each possible pair of alternatives (Zimmermann and Gutsche 1991; Triantaphyllou 2000):

- *strict preference*: one alternative is preferred to the other one
- *indifference*: two alternatives are regarded as equal
- *weak preference*: one alternative is either preferred to another or regarded as equal, depending on the circumstances

Depending on the criterion, one of six basic types of preference functions can be selected. As an example, for costs induced by a manufacturing process it might be strictly preferred to achieve lower costs compared to an alternative process, while zero costs would be the optimum (usual criterion). For the criterion processing time, an alternative might be strictly preferred to another only for processing times over a certain threshold p (u-shape criterion). Below that value (e.g. $p = 100$ s), alternatives are regarded as equal, because a further reduction of processing times may not feature additional benefits for the company. The preference functions are then applied to the deviations calculated in step one. As a result, *outranking relations* between all alternatives are received for each attribute (Behzadian et al. 2010). PI weights are then applied to the outranking relations in order to receive total positive and negative *outranking flows* for each alternative (compare Fig. 4.37d). Positive flows indicate the strength of an alternative, while negative flows reflect the weakness compared to all other alternatives (Brans et al. 1986). Finally, in step five the *net flow* as difference between positive and negative outranking flows is calculated for each alternative (compare Fig. 4.37e), providing a ranking of all alternatives from best (highest net flow) to worst (lowest net flow):

$$NF_{m_i} = OF_{m_i}^+ - OF_{m_i}^-$$ (4.33)

with:

NF_{m_i} net flow for measure m_i
$OF_{m_i}^+$ positive outranking flows for measure m_i
$OF_{m_i}^-$ negative outranking flows for measure m_i

A practical application of the presented methodology is described in Sect. 6.1.3.

4.5.4 Contributions to Resource Efficient Manufacturing

The improvement phase with its inherent steps and methods provides guidance to find and select improvement measures as alternative courses of action towards a more resource efficient manufacturing. An added value compared to existing approaches is particular been achieved regarding the following aspects:

- development of a *knowledge-based system (KBS)* for an automated and user-friendly identification of suitable improvement measures for a given improvement situation
- provision of a procedure for a *holistic model-based evaluation of measures and measure bundles,* allowing to identify target conflicts and problem shiftings
- presentation of two *situation-specific procedures to select improvement measures* for implementation under consideration of multi-criteria aspects using either weighted sums or PROMETHEE methods

4.6   Implications for Concept Operationalization

In order to respond to the research questions and requirements stated in Sect.4.1, an operationalization of the presented concept necessitates the provision of a software tool to support decision makers throughout all phase of the improvement procedure (addresses requirement R13). Consequently, the improvement procedure with its nine consecutive steps as described beforehand is translated into an implementable concept of a *decision support toolbox.* The three modules of the toolbox correspond to the phases of the improvement procedure: modeling, evaluation and improvement (compare Fig. 4.38). Data from real-world factories and value chains is used as input data to create virtual system models within the modeling module in accordance with objectives and targets followed by the decision maker (steps 1–3). The evaluation module is then used to apply methods for performance calculation, i.e. MEFA, VSM, LCC and LCA (step 4). The current system performance in terms

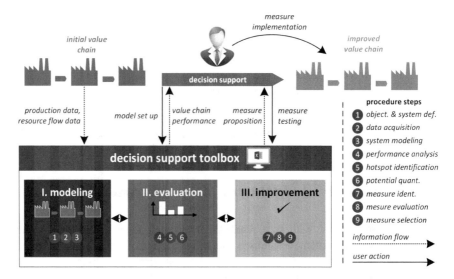

Fig. 4.38 Concept of a *decision support toolbox* for resource efficiency in manufacturing value chains

of technical, economic and environmental dimensions is used to derive indications about hotspots (step 5) and to quantify improvement potentials (step 6). In accordance with the objectives pursued, improvement measures are derived in the decision support module and proposed to the decision maker (step 7). Consequences arising from the implementation of measures can then be virtually tested in the evaluation module by applying changes to system structure and parameters (step 8) before using MCDA methods to select measures for implementation in the real value chain (step 9). In the following Chap. 5, the implementation of this architecture into a software demonstrator is described.

References

Ackoff RL (1989) From data to wisdom. J Appl Syst Anal 16(1):3–9

Allacker K, Mathieux F, Manfredi S et al (2014) Allocation solutions for secondary material production and end of life recovery: Proposals for product policy initiatives. Resour Conserv Recycl 88:1–12. https://doi.org/10.1016/j.resconrec.2014.03.016

Allwood JM (2012) Sustainable materials–with both eyes open. UIT Cambridge Ltd., Cambridge

Banks J, Carson JS, Nelson BL, Nicol D (2010) Discrete-event system simulation. Pearson, Upper Saddle River

Baumann H, Tillman A-M (2004) The Hitch Hiker's guide to LCA. Studentlitteratur, Lund

Bayerisches Landesamt für Umweltschutz (2003) Effiziente Energienutzung in der Galvanikindustrie. Augsburg

Behzadian M, Kazemzadeh RB, Albadvi A, Aghdasi M (2010) PROMETHEE: a comprehensive literature review on methodologies and applications. Eur J Oper Res 200:198–215. https://doi.org/10.1016/j.ejor.2009.01.021

Beier J (2017) Simulation approach towards energy flexible manufacturing systems. Springer International Publishing, Cham

Blume S, Herrmann C, Thiede S (2018) Increasing resource efficiency of manufacturing systems using a knowledge-based system. Procedia CIRP 69:236–241. https://doi.org/10.1016/j.procir.2017.11.126

Böge A, Böge W (2015) Handbuch Maschinenbau. Springer Vieweg, Wiesbaden

Bracht U, Geckler D, Wenzel S (2011) Digitale Fabrik - Methoden und Praxisbeispiele. Springer, Berlin

Brans JP, Vincke P, Mareschal B (1986) How to select and how to rank projects: the Promethee method. Eur J Oper Res 24(2):228–238. https://doi.org/10.1016/0377-2217(86)90044-5

Camp R, Tweet A (1994) Benchmarking applied to healthcare. J Qual Improv 20(5):229–238

Cerdas F, Egede P, Herrmann C (2018) LCA of electromobility. life cycle assessment. Springer International Publishing, Cham, pp 669–693

Cooper DR, Schindler PS (2014) Business research methods. McGraw-Hill, New York

Daenzer WF, Huber F (1994) Systems engineering: Methodik und Praxis. Verlag Industrielle Organisation, Zürich

Dehning P (2017) Steigerung der Energieeffizienz von Fabriken der Automobilproduktion. Springer Fachmedien, Wiesbaden

Dehning P, Thiede S, Mennenga M, Herrmann C (2017) Factors influencing the energy intensity of automotive manufacturing plants. J Clean Prod 142:2305–2314. https://doi.org/10.1016/j.jclepro.2016.11.046

Del Duce A, Egede P, Öhlschläger G, et al (2013) eLCAr – guidelines for the LCA of electric vehicles. European Union Seventh Framew Program

Despeisse M (2013) Sustainable manufacturing tactics and improvement methodology: a structured and systematic approach to identify improvement opportunities. Cranfield University

Deutsches Institut für Normung (2003) DIN 8580:2003–09 - Fertigungsverfahren - Begriffe, Einteilung

Deutsches Institut für Normung (2012) DIN EN 16231 - Energieeffizienz-Benchmarking-Methodik

Devoldere T, Dewulf W, Deprez W et al (2007) Improvement potential for energy consumption in discrete part production machines. Adv Life Cycle Eng Sustain Manuf Businesses 311–316. https://doi.org/10.1007/978-1-84628-935-4_54

Diaz N, Helu M, Jarvis A et al (2009) Strategies for minimum energy operation for precision machining. In: Proceedings of MTTRF 2009 annual meeting

Drumm C, Busch J, Dietrich W et al (2013) STRUCTese – Energy efficiency management for the process industry. Chem Eng Process Process Intensif 67:99–110. https://doi.org/10.1016/j.cep.2012.09.009

Duflou JR, Kellens K, Renaldi, et al (2012) Critical comparison of methods to determine the energy input for discrete manufacturing processes. CIRP Ann Manuf Technol 61(1):63–66. https://doi.org/10.1016/j.cirp.2012.03.025

Duflou JR, Kellens K, Renaldi DW (2011) Environmental performance of sheet metal working processes. Key Eng Mater 473:21–26. https://doi.org/10.4028/www.scientific.net/KEM.473.21

Dyckhoff H (2006) Produktionstheorie - Grundzüge industrieller Produktionswirtschaft, 5th edn. Springer, Berlin, Heidelberg, New York

Dyckhoff H, Spengler T (2010) Produktionswirtschaft, 3rd edn. Springer, Heidelberg

Eckey HF, Kosfeld R, Rengers M (2002) Multivariate Statistik Grundlagen - Methoden - Beispiele. Gabler, Wiesbaden

Ecoinvent Centre - Swiss Centre for Life Cycle Inventories (2018) ecoinvent database v3.5. https://www.ecoinvent.org/. Accessed 9 Aug 2019

Erlach K (2010) Wertstromdesign. Springer, Berlin

European Commission (2018) Reference documents under the IPPC Directive and the IED. https://eippcb.jrc.ec.europa.eu/reference/. Accessed 10 Aug 2019

European Union (2012) Directive 2012/19/EU of the European Parliament and of the Council of 4 July 2012 on waste electrical and electronic equipment (WEEE)

Fayyad U, Piatetsky-Shapiro G, Smyth P (1996) From data mining to knowledge discovery in databases. AI Mag 17(3):37–54

Flick D, Ji L, Dehning P et al (2017) Energy efficiency evaluation of manufacturing systems by considering relevant influencing factors. Procedia CIRP 63:586–591. https://doi.org/10.1016/j.procir.2017.03.097

Ghadimi P, Li W, Kara S, Herrmann C (2014) Integrated material and energy flow analysis towards energy efficient manufacturing. Procedia CIRP 15:117–122. https://doi.org/10.1016/j.procir.2014.06.010

Gordon TJ (1994) Cross-impact method. https://citeseerx.ist.psu.edu/viewdoc/download?doi=10.1.1.202.7337&rep=rep1&type=pdf. Accessed 11 July 2020

Götze U (2010) Kostenrechnung und Kostenmanagement. Springer, Berlin

Graedel TE, Allwood J, Birat J-P et al (2011) Recycling rates of metals – a status report. UNEP, Nairobi

Günther H-O, Tempelmeier H (2005) Produktion und Logistik. Springer, Berlin

Gutowski T, Dahmus J, Thiriez A (2006) Electrical energy requirements for manufacturing processes. In: 13th CIRP international conference on life cycle engineering, pp 623–628

Gutowski T, Dahmus J, Thiriez A et al (2007) A thermodynamic characterization of manufacturing processes. In: Proceedings of the 2007 IEEE international symposium on electronics and the environment. IEEE, pp 137–142

Gutowski TG, Allwood JM, Herrmann C, Sahni S (2013a) A global assessment of manufacturing: economic development, energy use, carbon emissions, and the potential for energy efficiency and materials recycling. Annu Rev Environ Resour 38(1):81–106. https://doi.org/10.1146/annurev-environ-041112-110510

Gutowski TG, Sahni S, Allwood JM, et al (2013b) The energy required to produce materials: constraints on energy-intensity improvements, parameters of demand. Philos Trans R Soc A Math Phys Eng Sci 371. https://doi.org/10.1098/rsta.2012.0003

Haag H (2013) Eine Methodik zur modellbasierten Planung und Bewertung der Energieeffizienz in der Produktion. Fraunhofer Verlag, Stuttgart

Hauschild MZ (2015) Better – but is it good enough? On the need to consider both eco-efficiency and eco-effectiveness to gauge industrial sustainability. Procedia CIRP 29:1–7. https://doi.org/10.1016/j.procir.2015.02.126

Heinemann T (2016) Energy and resource efficiency in aluminium die casting. Springer International Publishing, Cham

Herrmann C, Thiede S (2009) Process chain simulation to foster energy efficiency in manufacturing. CIRP J Manuf Sci Technol 1:221–229. https://doi.org/10.1016/j.cirpj.2009.06.005

Herrmann C, Zein A, Winter M, Thiede S (2010) Procedures and tools for metering energy consumption of machine tools. In: Manuf 2010 – 3rd Int Sci Conf with Expert Particip

Hesselbach J (2012) Energie- und klimaeffiziente Produktion. Vieweg+Teubner Verlag, Wiesbaden

Ilgin MA, Gupta SM (2010) Environmentally conscious manufacturing and product recovery (ECMPRO): A review of the state of the art. J Environ Manage 91(3):563–591. https://doi.org/10.1016/j.jenvman.2009.09.037

International Organization for Standardization (2006a) ISO 14040 – environmental management – life cycle assessment – principles and framework

International Organization for Standardization (2006b) ISO 14044 – Environmental management – life cycle assessment – requirements and guidelines

Jawahir IS, Dillon OW, Rouch KE et al (2006) Total life-cycle considerations in product design for sustainability: a framework for comprehensive evaluation. In: Proceedings of 10th Internatioanl Research Conference on TMT 2006, Barcelona-Lloret Mar, Spain, 11–15 Sept 2006

Kalla D, Twomey J, Overcash M (2009) MC 1 brake forming process – unit process life cycle inventory. https://cratel.wichita.edu/uplci/wp-content/uploads/2009/08/MC1-brake-for ming-6-9-2009_Final.doc

Kara S, Bogdanski G, Li W (2011) Electricity metering and monitoring in manufacturing systems. In: Glocalized solutions for sustainability in manufacturing – proceedings of the 18th CIRP international conference on life cycle engineering

Keichel C, Volta D, Sankol B (2015) Optimierung ist endlich - Neue Methodik zur systematischen Bewertung und Verbesserung der Energieeffizienz von Produktionsanlagen und -prozessen. In: Biedermann H, Vorbach S (eds) Innovation und Nachhaltigkeit: Strategisch-operatives Energie- und Ressourcenmanagement. Rainer Hampp Verlag, München

Kim S, Hwang T, Lee K (1997) Allocation for cascade recycling system. Int J Life Cycle Assess 2(4):217–222

Klöpffer W (1997) In defense of the cumulative energy demand. Int J Life Cycle Assess 2(2):61. https://doi.org/10.1007/BF02978754

Kreitlein S, Scholz M, Franke J (2017) The automated evaluation of the energy efficiency for machining applications based on the least energy demand. Procedia CIRP 61:404–409

Kruschwitz L (2014) Investitionsrechnung. Walter de Gruyter GmbH & Co KG, Berlin

Kwon O, Yoo K, Suh E (2005) UbiDSS: a proactive intelligent decision support system as an expert system deploying ubiquitous computing technologies. Expert Syst Appl 28(1):149–161. https://doi.org/10.1016/j.eswa.2004.08.007

Lässig J, Schütte T, Riesner W (2016) Energieeffizienz-Benchmark Industrie. Springer Fachmedien, Wiesbaden

Li W, Zein A, Kara S, Herrmann C (2011) An investigation into fixed energy consumption of machine tools. Glocalized Solut Sustain Manuf 268–273. https://doi.org/10.1007/978-3-642-196 92-8

Löfgren B (2009) Capturing the life cycle environmental performance of a company's manufacturing system. Chalmers University of Technology

Luger T (2010) Referenzprozessbasierte Gestaltung und Bewertung von reverse supply chains. Vulkan, Essen

Mahamud R, Li W, Kara S (2017) Energy characterisation and benchmarking of factories. CIRP Ann 66(1):457–460. https://doi.org/10.1016/j.cirp.2017.04.010

Mardan N, Klahr R (2012) Combining optimisation and simulation in an energy systems analysis of a Swedish iron foundry. Energy 44(1):410–419. https://doi.org/10.1016/j.energy.2012.06.014

Millot C, Fillot LA, Lame O et al (2015) Assessment of polyamide-6 crystallinity by DSC: Temperature dependence of the melting enthalpy. J Therm Anal Calorim 122(1):307–314. https://doi.org/10.1007/s10973-015-4670-5

Moen R, Norman C (2009) Evolution of the PDCA cycle. Society

Nicholson A, Olivetti E, Gregory J et al (2009) End of life allocation methods: open loop recycling impacts on robustness of material selection decisions 2009. In: IEEE Internatioanl Symposium Sustainable Systems and Technology, pp 1–6. https://doi.org/10.1109/ISSST.2009.5156769

Posselt G (2016) Towards energy transparent factories. Springer International Publishing, Cham

Posselt G, Fischer J, Heinemann T et al (2014) Extending energy value stream models by the TBS dimension – applied on a multi product process chain in the railway industry. Procedia CIRP 15:80–85. https://doi.org/10.1016/j.procir.2014.06.067

Prakash S, Dehoust G, Gsell M, Schleicher T (2015) Einfluss der Nutzungsdauer von Produkten auf ihre Umweltwirkung: Schaffung einer Informationsgrundlage und Entwicklung von Strategien gegen „Obsoleszenz". Umweltbundesamt, Dessau-Roßlau

Puppe F (1991) Einführung in Expertensysteme. Springer, Berlin

Rahimifard S, Seow Y, Childs T (2010) Minimising embodied product energy to support energy efficient manufacturing. CIRP Ann Manuf Technol 59:25–28. https://doi.org/10.1016/j.cirp.2010.03.048

Reinhardt S (2013) Bewertung der Ressourceneffizienz in der Fertigung. Herbert Utz Verlag, München

Rockström J, Steffen WL, Noone K, Al E (2009) Planetary Boundaries: Exploring the safe operating space for humanity. Ecol Soc 14(2):81–87. https://doi.org/10.1007/s13398-014-0173-7.2

Saltelli A, Ratto M, Andres T et al (2008) Global sensitivity analysis: the primer. Wiley, Chichester

Sapsford R, Jupp V (2006) Data collection and analysis 2. SAGE Publications, Thousand Oaks

Simeone A, Luo Y, Woolley E et al (2016) A decision support system for waste heat recovery in manufacturing. CIRP Ann Manuf Technol 65(1):21–24. https://doi.org/10.1016/j.cirp.2016.04.034

Simon S (2006) Benchmarking im Werkzeugmaschinenbau. Technische Universität Darmstadt, Ein Beitrag zur wettbewerblichen Produktentwicklung

Steinhaus H (1956) Sur la division des corps materiels en parties. Bull Polish Acad Sci 4(3):801–804

Tan YS, Tjandra TB, Song B (2015) Energy efficiency benchmarking methodology for mass and high-mix low-volume productions. Procedia CIRP 29:120–125. https://doi.org/10.1016/j.procir.2015.02.200

Teiwes H, Blume S, Herrmann C et al (2018) Energy load profile analysis on machine level. Procedia CIRP 69:271–276. https://doi.org/10.1016/j.procir.2017.11.073

Thiede S (2012) Energy efficiency in manufacturing systems. Springer, Berlin

Thiede S, Li W, Kara S, Herrmann C (2016) Integrated analysis of energy, material and time flows in manufacturing systems. Procedia CIRP 48:200–205. https://doi.org/10.1016/j.procir.2016.03.248

Thiriez A, Gutowski T (2006) An environmental analysis of injection molding. In: Proceedings of the 2006 IEEE international symposium on electronics and the environment, pp 195–200

Triantaphyllou E (2000) Multi-criteria decision making methods: a comparative study. Springer Science+Business Media, Dordrecht

Turban E, Aronson JE, Liang T-P (2007) Decision support systems and intelligent systems. Prentice Hall, Upper Saddle River

Ude J (2010) Entscheidungsunterstützung für die Konfiguration globaler Wertschöpfungsnetzwerke. Shaker Verlag, Aachen

United Nations (1998) Kyoto protocol to the united nations framework convention on climate change

United Nations (2015) Paris Agreement

United Nations Environment Programme (2011) Global guidance principles for life cycle assessment databases. Nairobi

United Nations Industrial Development Organization (2014) Industrial energy efficiency project – benchmarking report for the iron and steel sector

Verband Deutscher Maschinen- und Anlagenbau e.V. (2015) VDMA 34179 - Messvorschrift zur Bestimmung des Energie- und Medienbedarfs von Werkzeugmaschinen in der Serienfertigung. VDMA, Frankfurt am Main

Verein Deutscher Ingenieure (2015) VDI 4600:2012-01 - Kumulierter Energieaufwand (KEA); Begriffe, Berechnungsmethoden. VDI, Düsseld

Volta D (2014) Das Physikalische Optimum als Basis von Systematiken zur Steigerung der Energie- und Stoffeffizienz von Produktionsprozessen. Technische Universität Clausthal

Walther G (2010) Nachhaltige Wertschöpfungsnetzwerke. Gabler, Wiesbaden

Winter M (2016) Eco-efficiency of grinding processes and systems. Springer International Publishing, Cham

Winter M, Li W, Kara S, Herrmann C (2014) Determining optimal process parameters to increase the eco-efficiency of grinding processes. J Clean Prod 66:644–654. https://doi.org/10.1016/j.jclepro.2013.10.031

Womack JP, Jones DT (1996) Lean thinking – banish waste and create wealth in your corporation. The Free Press, New York

Yoon HS, Lee JY, Kim HS et al (2014) A comparison of energy consumption in bulk forming, subtractive, and additive processes: Review and case study. Int J Precis Eng Manuf Green Technol 1(3):261–279. https://doi.org/10.1007/s40684-014-0033-0

Yuan C, Zhai Q, Dornfeld D (2012) A three dimensional system approach for environmentally sustainable manufacturing. CIRP Ann Manuf Technol 61(1):39–42. https://doi.org/10.1016/j.cirp.2012.03.105

Zein A (2013) Transition towards energy efficient machine tools. Springer, Berlin

Zhou M, Pan Y, Chen Z, Yang W (2013) Optimizing green production strategies: an integrated approach. Comput Ind Eng 65(3):517–528. https://doi.org/10.1016/j.cie.2013.02.020

Zimmermann H-J, Gutsche L (1991) Multi-criteria analyse. Springer, Berlin

Chapter 5
Concept Implementation

This chapter deals with the software implementation of the new concept as *decision support toolbox*, starting with the description of requirements for the tool in Sect. 5.1. Subsequently, the implementation into a selected software environment is described in Sect. 5.2 for all three modules – modeling, evaluation and improvement. In order to allow for a high applicability, the intended workflow from a user perspective is outlined in Sect. 5.3. The chapter closes with a description of the applied verification and validation procedures to ensure the correctness of results generated with the tool in Sect. 5.4.

5.1 Requirements

The requirements for implementation into a software demonstrator are highly related to the requirements for the concept itself (compare Sect. 4.1). Accordingly, the tool should allow.

- to build up detailed system models founding on the MEFA method
- to apply LPC, easing the interpretation of dynamic resource demand profiles
- to conduct performance analyses and calculate meaningful PIs based on VSM, LCA and LCC methods to cover technical, environmental and economic evaluation dimensions
- to set up and apply a knowledge-based system for measure identification
- to apply MCDA methods for measure selection
- to store and manage underlying model input and output data
- to adequately visualize the results

S. A. Blume, *Resource Efficiency in Manufacturing Value Chains*,
Sustainable Production, Life Cycle Engineering and Management,
https://doi.org/10.1007/978-3-030-51894-3_5

Further, requirement R13 demands for a user-friendly and seamless software implementation and a high degree of automation. As users may belong to different groups in terms of qualification or interest, the tool should offer customized user modes. To reduce complexity and achieve a seamless implementation, the utilization of multiple software environments should rather be avoided. According to R14, standard software is preferred due to lower barriers for utilization in terms of costs or user knowledge as well as easier distribution of results. Thus, *MS Excel* is selected for implementation as it is one of the most widely used office software tools and most users are familiar with its basic functions. With *Visual Basic for Applications* it provides an integrated programming language to build user-defined functions and to automate processes. In order to reduce modeling and utilization efforts (R15), a significant degree of automation is regarded as essential feature together with a pre-definition of manufacturing system elements. As a good scalability of the concept to cover use cases with varying level of detail and complexity is intended, a modular tool architecture is favored. Hence, users may only apply those modules which are needed for their specific use case and application scenario. Further, users shall be enabled to make adaptions and extensions to suit other use cases with relatively low efforts (R17). As manipulation of user interfaces and program code is relatively simple in *MS Excel* compared to other software environments, software experts or programmers are not necessarily required for minor tool adaptions.

5.2   Implementation as Software Demonstrator

In accordance with the requirements presented ahead, the concept is implemented as a modular *decision support toolbox* in *MS Excel* by developing seven connectable file types:

- **Data interpretation files (DI files)** allow to apply the LPC methodology in order to reduce manual efforts of resource demand profile interpretation (compare Sect. 4.3.2.2).
- **Single factory files (SF files)** are used to build up detailed system models of manufacturing systems. *SF files* constitute the core of the toolbox, as they comprise information about products, resource flows, machines, processes and factories as well as structural relations between the elements. All elements and their attributes are pre-defined and stored in model galleries, i.e. the user rather conducts an arrangement or configuration, avoiding an effortful modeling from scratch. Assessment and evaluation of the system performance (compare Sect. 4.4) is mainly performed in an automated manner. To form a value chain, multiple *SF files* can be physically connected. Further, *SF files* can also be used to model other life cycle stages.
- **Eco impact files** contain information about environmental consequences resulting from resource utilization, i.e. environmental impact factors. These are needed for environmental PI calculation (compare Sect. 4.4.1.3).

- **Benchmarking files (BM files)** contain information about reference values for specific manufacturing processes, allowing to benchmark hotspot processes and quantify their performance gaps (compare Sect. 4.4.3).
- **Value chain files (VC files)** contain detailed as well as aggregated performance indicator values from SF files to conduct a performance assessment on value chain level and along the product life cycle.
- **KBS files** contain a knowledge-based system to identify suitable improvement measures from a knowledge base. Potentially fitting measures are proposed to the user (compare Sect. 4.5.1).
- **MCDA files** support the application of MCDA methods such as PROMETHEE to select one or a bundle of measures for implementation taking into account improvement potential as well as user targets (compare Sect. 4.5.3).

As depicted in Fig. 5.1, the files relate to different phases and steps of the underlying improvement procedure. At least one *single factory file* is mandatory to perform an analysis, whereas the use of all other file types is optional and depends on the objectives followed. For instance, *eco impact files* are only required for analyses covering the environmental target dimension, while a *KBS file* is only required to automatically receive proposals for possible improvement measures. In the following sections, the different file types and their interactions are explained in detail.

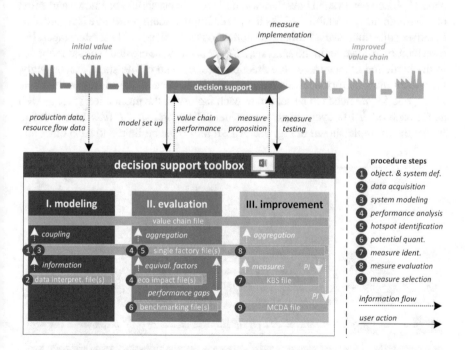

Fig. 5.1 Concept implementation in *MS Excel* and interaction between user, toolbox and files

5.2.1 Data Interpretation File

DI files can be used for data interpretation in *step 2: data acquisition* to reduce efforts for manual metering data analysis (compare Sect. 4.3.2.2) by application of the LPC methodology. As result, state-based resource demand levels, in this case average electrical loads of machines in different machine states, can be calculated. Metering data must be available in a format readable by *MS Excel,* featuring time stamp and demand value for each second of the metering period. An import and analysis of such time series can then be performed automatically through a user query. As additional information, the processing time *PT* for the product of interest is needed in order to reach a higher validity of results. An exemplary application is shown in Sect. 6.2.1.

5.2.2 Single Factory File

SF files form the core of the toolbox, containing manufacturing system models from process to factory level. Before starting to model a real-world system using *SF files*, steps 1 and 2 of the improvement procedure need to be carried out. Consequently, general objectives, system boundaries and PIs of interest should be known and a first data basis should be established. As the whole improvement procedure is regarded as iterative, adjustments and further data collections are still possible at later stages. For each factory to consider in the analysis, a *SF file* needs to be created. The same applies to the modeling of raw material extraction, use and end of life stage. Accordingly, for a local intra-factory analysis only one *SF file* is needed. To address a life cycle scope, one *SF file* needs to be set up for each factory in the production stage as well as for each other life cycle stage to consider. Therefore, six *SF files* are needed to model the example shown in Fig. 5.2. All *SF files* must be linked to each other and

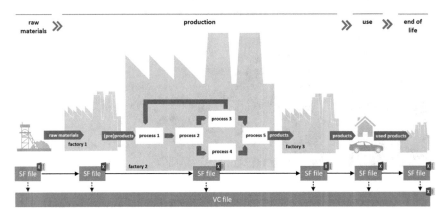

Fig. 5.2 *SF files* required for an exemplary life cycle analysis with four factories considered in the production stage

can then be coupled with a *VC file* to receive PIs on higher system levels (compare Sect. 5.2.4).

In order to allow for a meaningful application of the toolbox for different stakeholders and in different situations, three user modes are implemented into the *SF file* template:

- The *expert mode* activates the whole range of toolbox functionalities. This mode is mainly used for initial modeling and adjustment of the model structure.
- The *simple mode* does activate all basic functions, but only allows for parameter modifications in existing models to check the consequences of measure implementation, while structural model changes are not supported.
- The *presentation mode* allows to show results of the conducted analyses. Changes to models cannot be applied and all toolbars are hidden to optimize visualization.

The actual system modeling usually starts with the task to define a new factory, i.e. to set general parameters on factory level (compare Table 4.8 in Sect. 4.3.3.2). Subsequently, the factory elements of the factory need to be created according to the manufacturing system structure. Figure 5.3 shows the corresponding worksheet, which is divided into three areas:

(a) Figure 5.3a deals with the definition of processes according to the intra-factory process chain to be analyzed. Here, all manufacturing processes need to be created, defining main group and process type (compare Table 2.1 in Sect. 2.1.2). This information is a prerequisite for *step 6: potential quantification* and *step 7: measure identification* of the improvement procedure.

(b) In Fig. 5.3b, corresponding machines for created processes need to be set up. In order to conduct a process, at least one machine must be assigned. Machines can be parametrized after creation by setting machine related parameters (compare Table 4.7 in Sect. 4.3.3.2). To allow for a good usability and to guide the user through the steps to be performed, interactive user forms are used for both creation and parameterization of most factory elements. Automated plausibility checks are integrated wherever meaningful, e.g. percentage values such as

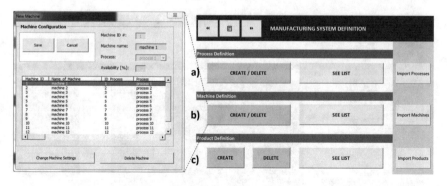

Fig. 5.3 Manufacturing system definition in *SF file* (screenshot)

machine availability must be assigned a value from 0 to 100. Drop down cells with pre-defined entries are used in many cases in order to prevent incorrect inputs by spelling errors or selection of non-existing elements.

(c) In Fig. 5.3c, new product types can be created. Besides product related parameters such as product mass (compare Table 4.3 in Sect. 4.3.3.2) this step also entails the definition of the in-house process chain structure, i.e. the sequence of process steps to be conducted for a specific product. Each created product type is constituted by an individual order of processes and thus an individual factory structure matrix (compare Sect. 4.3.3.1).

As soon as the elements of a factory are created and machines and products are parametrized, process parametrization is a next step to be performed. For each product type, one new worksheet is created per process step in order to set resource and process related parameters. A screenshot of such a sheet is presented in Fig. 5.4, whereby the main areas for user inputs are highlighted. In Fig. 5.4a, process related parameters like processing time, batch size or quality rate are defined (compare Table 4.6 in Sect. 4.3.3.2). In Fig. 5.4b, resource inputs need to be quantified, distinguishing between energies, materials and labor (compare Table 4.4 in Sect. 4.3.3.2). In Fig. 5.4c, corresponding outputs such as products, waste and emissions are quantified. Cost factors for all resources are defined on a separate sheet, as they are assumed to be identical for all processes within one factory.

After process parameterization, the setup of a *SF file* is basically finished. If the analysis scope goes beyond a single factory, several *SF files* can be coupled and linked according to the connecting product flow. Therefore, in each *SF file* a successor, i.e. the downstream *SF file*, needs to be selected except for the last factory. Every *SF file* requires exactly one successor, while it can have multiple predecessors (n:1 connection). Accordingly, continuous and converging material flows can be modelled on value chain level, whereas diverging or rearranging flows are excluded for reasons of simplicity (compare Sect. 2.1.2). As soon as all required elements are created and parameterized, a performance analysis can be automatically carried out for a selected product. Results of the performance analysis are then displayed on

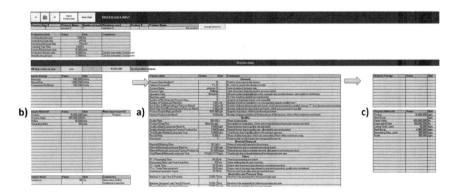

Fig. 5.4 Process parametrization sheet (screenshot)

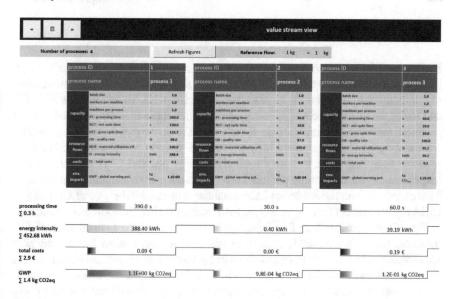

		1			2			3
process ID		1	process ID		2	process ID		3
process name		process 1	process name		process 2	process name		process 3
capacity	batch size	- 1.0	capacity	batch size	- 1.0	capacity	batch size	- 1.0
	workers per machine	- 1.0		workers per machine	- 1.0		workers per machine	- 1.0
	machines per process	- 1.0		machines per process	- 1.0		machines per process	- 1.0
	PT - processing time	s 390.0		PT - processing time	s 30.0		PT - processing time	s 60.0
	NCT - net cycle time	s 130.0		NCT - net cycle time	s 10.0		NCT - net cycle time	s 20.0
	GCT - gross cycle time	s 132.7		GCT - gross cycle time	s 10.3		GCT - gross cycle time	s 20.0
resource flows	QR - quality rate	% 98.0	resource flows	QR - quality rate	% 97.0	resource flows	QR - quality rate	% 100.0
	MUE - material utilization eff.	% 100.0		MUE - material utilization eff.	% 100.0		MUE - material utilization eff.	% 95.2
	EI - energy intensity	kWh 388.4		EI - energy intensity	kWh 0.4		EI - energy intensity	kWh 39.2
costs	TC - total costs	€ 0.1	costs	TC - total costs	€ 0.0	costs	TC - total costs	€ 0.2
env. impacts	GWP - global warming pot.	kg CO2eq 1.1E+00	env. impacts	GWP - global warming pot.	kg CO2eq 9.8E-04	env. impacts	GWP - global warming pot.	kg CO2eq 1.2E-01

processing time $\sum 0.3$ h	390.0 s	30.0 s	60.0 s
energy intensity $\sum 452.68$ kWh	388.40 kWh	0.40 kWh	39.19 kWh
total costs $\sum 2.9$ €	0.09 €	0.00 €	0.19 €
GWP $\sum 1.4$ kg CO2eq	1.1E+00 kg CO2eq	9.8E-04 kg CO2eq	1.2E-01 kg CO2eq

Fig. 5.5 Value stream view of an exVSM map for an exemplary product (screenshot)

several worksheets. An example for a value stream view as part of an exVSM map (compare Sect. 4.4.1.4) is presented in Fig. 5.5.

SF files are also used to support *step 5: hotspot identification* of the improvement procedure by conducting sensitivity analyses (compare Sect. 4.4.2.2). In order to identify critical parameters and processes of the system, a *variation factor* must be defined by the user (compare Fig. 5.6a). All changeable input parameters of the model are then automatically multiplied with this factor. For some parameters this may mean an improvement (e.g. increased quality rate), for others it may mean the opposite (e.g. increased energy intensity). Hence, sensitivity analyses should be conducted into both positive and negative direction by using positive and negative variation factors. The resulting relative PI changes are displayed as graphs to enable a quick identification of the most relevant parameters (compare Fig. 5.6b).

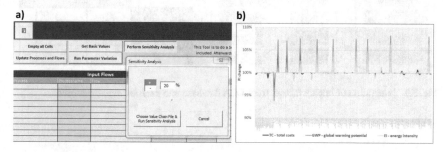

Fig. 5.6 Exemplary sensitivity analysis with selection of variation factor (**a**) and resulting PI changes (**b**) (screenshot)

5.2.3 Eco Impact File

Eco impact files contain information about environmental impacts of resource flows, i.e. of process inputs and outputs. *SF files* must be linked to an *Eco impact file* in order to calculate environmental PIs as described in Sect. 4.4.1.3. To increase the usability and lower the effort for environmental assessments, more than 300 (semi-finished) products, materials and energies are already available in the *Eco impact file* template, representing typical resources which are used in discrete manufacturing. For each resource flow, impact factors for cumulative energy demand *CED*, climate change (expressed as global warming potential *GWP*) and other environmental impacts have been extracted from the *ecoinvent* LCA database. The relation between LCI data and LCIA results (impact factors) as well as their integration into the toolbox is illustrated in Fig. 5.7. As indicated, an *Eco impact file* can be understood as local offline LCA database, which is seamlessly integrated into the toolbox. Linked *Eco impact files* are automatically used for PI calculations in *SF files*, while expert knowledge for the utilization of additional LCA software is not required. Yet, the toolbox user ought to select adequate datasets from the *Eco impact file*, which fit well the resource flows used in the *SF file*. Resource characteristics such as material composition (e.g. share of different metals in a steel alloy), related production process or process route (e.g. blast furnace or electric arc steelmaking) and source country or world region (e.g. steel sourced from Europe or China) should be taken into account.

As a consequence, from the simplified approach to conduct environmental analyses, results must not be treated as full LCA complying with ISO 14000 standards as described in Sect. 2.4.5. They shall rather give first indications about the environmental relevance of different resources and processes involved in product creation. Further, they can also be used as basis for a more detailed LCA study in a next step.

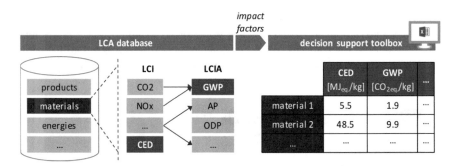

Fig. 5.7 Integration of impact factors from LCA database into *decision support toolbox*

5.2.4 Value Chain File

VC files are mainly needed to carry out *step 4: performance analysis* and *step 8: measure evaluation* for scopes beyond a factory's gates. For this purpose, they can import performance indicator results from coupled *SF files* and compile aggregated PIs for higher system levels (compare Fig. 4.20 in Sect. 4.4.1). PIs can be calculated and visualized for selected *SF files* of interest as described in Sect. 4.4.1.4. Figure 5.8 shows the cockpit view of an exVSM map with individual PIs for every imported *SF file* (compare Fig. 5.8a) and aggregated PIs for the entire value chain (compare Fig. 5.8b). Individual contributions of *SF files* to PIs are visualized by means of diagrams (compare Fig. 5.8c). PIs displayed in the cockpit view are customizable and can differ from the PIs displayed in the value stream view, depending on the requirements of the analysis.

A function to save the current set of PI values is integrated, allowing for a testing and comparison of different improvement measures in *step 8: measure evaluation*. Changes can be applied to one or several *SF files*, while arising consequences can be directly assessed in the results visualization of a coupled *VC file*.

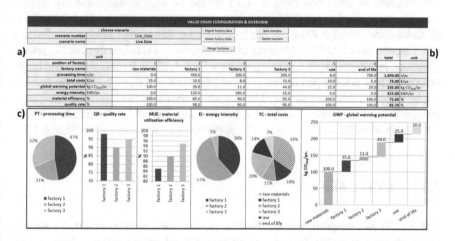

Fig. 5.8 Cockpit view of exVSM map for an exemplary value chain (screenshot)

5.2.5 Benchmarking File

BM files can be created to support *step 6: potential quantification* in order to quantify performance gaps (compare Sect. 4.4.3). For this reason, PIs can be imported from *SF files* into a corresponding *BM file* to compare the performance of a critical process with best available technology *BAT* and physical optimum *PhO* reference values. In each process specific *BM file*, individual processing parameters are indicated for each single *BAT* value, so the user can select the one with the highest similarity to the current use case. In contrast, *PhO* values are expressed by equations, as they can mathematically be traced back to product properties and processing parameters. Exemplary applications are shown in Sects. 6.1.2 and 6.2.2.

5.2.6 KBS File

The *KBS file* template contains a knowledge-based system to apply *step 7: measure identification* as described in Sect. 4.5.1. It can be used to automatically receive measure proposals that are tailored to the current improvement situation. Therefore, all three elements of a KBS are implemented, i.e. knowledge database, inference machine and user interface. Measure characteristics are checked against the results generated in the sensitivity analysis (compare Sect. 4.4.2.2). Required data and information such as critical parameters, critical processes and other process parameters can be imported from any SF file. The measure retrieval by the inference machine is then performed automatically. Results, i.e. measures with differing matching levels, are visualized in the user interface. The application is demonstrated in Sect. 6.1.3.

5.2.7 MCDA File

The *MCDA file* template can be used to support the selection of one or several measures for implementation in *step 9: measure selection* (compare Sect. 4.5.3). Therefore, the PROMETHEE method is used to cope with competing targets and potential target conflicts of analyzed measures. The user is guided through the consecutive steps of the method, i.e. weighting of PIs, assignment of preference functions, calculation of outranking relations and outranking flows until final aggregation and ranking of alternatives according to the net flows (compare Sect. 4.5.3.2). An exemplary application is described in Sect. 6.1.3.

5.3 Application Procedure from User Perspective

The toolbox architecture follows a modular approach to allow for a flexible and use case specific concept application. Depending on the aspects to consider in an analysis, some procedure steps might be skipped or performed without explicit tool support. Accordingly, some toolbox files might not be needed in every use case. Figure 5.9 provides an application procedure, involving relevant decision points for modeling and application (compare requirement R12 in Sect. 4.1).

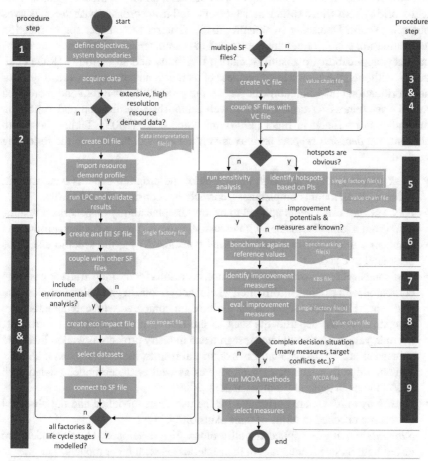

DI: data interpretation KBS: knowledge-based system MCDA: multi-criteria decision analysis
PI: performance indicator SF: single factory VC: value chain

Fig. 5.9 Application procedure for toolbox application with respect to needed toolbox files

5.4 Verification and Validation

Incorrect results of an analysis can have significant negative impacts, if they are used for decision making. Verification and validation techniques should be applied in order to ensure the correctness of models as well as generated results. Verification, on the one hand, deals with the formal correctness of a model and can for instance be assured by checking model syntax and consistency (Verein Deutscher Ingenieure 2018). Accordingly, it ensures a correct implementation of a conceptual model into a software model. Validation investigates, on the other hand, if a model represents the real-world system with a sufficient degree of detail in accordance with the objectives pursued (Verein Deutscher Ingenieure 2018). Criteria to consider for verification and validation comprise but are not limited to completeness, consistency, accuracy, currency, applicability, plausibility, clarity, feasibility and accessibility (Rabe et al. 2008). Although the complete correctness of a model can never be proven, a systematic verification and validation can reduce the probability of errors and increase a model's usefulness. Overviews about well-established verification and validation methods can be found in Balci (1998) and Rabe et al. (2008). Taking those into account, the *decision support toolbox* is verified and validated using the following methods:

- *Desk checking*, i.e. a thorough inspection of the program code, is conducted in order to check model completeness, correctness, consistency and clarity.
- *Face validity checking* is applied to ensure the plausibility of generated results, e.g. through joint discussions with industrial experts (process owners, production managers, managing directors, external consultants) for the specific use cases presented in Chap. 6.
- A *model output comparison* is carried out, i.e. results for the industrial use cases are compared with existing information wherever possible. In particular, technical and cost related PIs generated with the toolbox are compared with calculations from company internal departments such as production management or controlling. Further, values from conducted research and industry projects as well as literature values are taken into account. Environmental results, in contrast, cannot be fully validated due to a lack of existing studies as well as uncertainties related with LCA studies in general (Finnveden et al. 2009).
- *Sensitivity analyses* are performed for the use cases modelled and the observed effects are checked in terms of their plausibility.
- *Submodel testing* is applied for all toolbox files developed, i.e. all models are tested both independently and in their interaction with other models.

As a result, the toolbox is regarded to be thoroughly verified and validated. However, further validations such as face validity checking and model output comparisons are always recommended when applying the toolbox to a new use case in order to identify errors in input data or model setup.

5.5 Contributions to Resource Efficient Manufacturing

The implementation of the *decision support toolbox* as software tool provides valuable contributions to a more resource efficient manufacturing. In particular, this applies for the following aspects:

- provision of a *well-advanced software prototype* that is successfully validated through the application on different industrial use cases (compare Chap. 6)
- *reduction of efforts for manufacturing system modeling and improvement* due to provision of pre-defined model elements and model galleries as well as decision support with a high degree of automation

References

Balci O (1998) Verification, validation, and testing. In: Banks J (ed) Handbook of simulation. Wiley, Atlanta

Finnveden G, Hauschild MZ, Ekvall T et al (2009) Recent developments in life cycle assessment. J Environ Manage 91(1):1–21. https://doi.org/10.1016/j.jenvman.2009.06.018

Rabe M, Spieckermann S, Wenzel S (2008) Verifikation und Validierung für die Simulation in Produktion und Logistik. Springer, Berlin

Verein Deutscher Ingenieure (2018) VDI 3633 – Simulation of systems in materials handling, logistics and production – Terms and definitions. VDI, Düsseldorf

Chapter 6
Concept Application in the Metal Mechanic Industry

This chapter focuses on the application of the new concept to real-world industrial use cases with the objective to increase the resource efficiency of involved factories. In Sect. 6.1, a piston rod manufacturing is presented as first use case, involving multiple factories along a value chain from the metal mechanic sector. This case study has partly been carried out within the frame of the *MEMAN (Integral Material and Energy Flow Management in Manufacturing Metal Mechanic Sector)* project, which has received funding from the European Union under the research program Horizon 2020. The case study is used to demonstrate the whole potential of the developed concept for a holistic assessment of entire value chains, taking into account the whole product life cycle. In Sect. 6.2, an axle journal manufacturing as second case study is presented, focusing on a single factory perspective and selected resource flows.

6.1 Case Study 1: Piston Rod Manufacturing

6.1.1 Phase I: Modeling

The first case study deals with the manufacturing of medium sized piston rods made from tempered steel. They are widely used in hydraulic cylinders of construction vehicles like caterpillars or cranes. In *phase I: modeling*, a virtual system model is created using the developed *decision support toolbox*. Figure 6.1 highlights the challenge to set up a model for the product under assessment, which covers the whole product life cycle and contains three factories that transform steel billets into piston rods during the production stage.

S. A. Blume, *Resource Efficiency in Manufacturing Value Chains*,
Sustainable Production, Life Cycle Engineering and Management,
https://doi.org/10.1007/978-3-030-51894-3_6

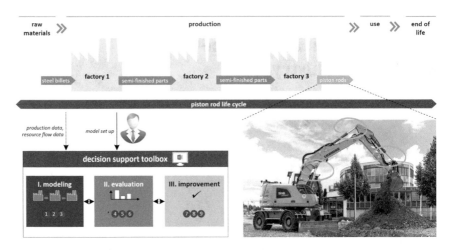

Fig. 6.1 Value chain and life cycle of piston rods to be used in construction machines, image taken from Liebherr Hydraulikbagger GmbH (2019)

As a first step, objectives and system boundaries need to be defined (compare Sect. 4.3.1). Piston rods are selected as case product type, as they constitute a key product in terms of production volumes and revenues for the companies involved in the analyzed value chain. At the same time, significant reject rates in *factory 2* are already known. Accordingly, high improvement potentials can be assumed (compare Fig. 6.2).

As main objective of the analysis a global and holistic system assessment is aspired, aiming at a system improvement in terms of technical, economic and environmental performances. Due to the holistic focus, modeling needs to consider all life cycle stages from raw material extraction until end of life (compare Fig. 6.3). In terms of manufacturing system levels, the entire value chain is analyzed within

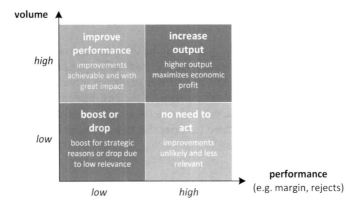

Fig. 6.2 Selection of piston rods with high volume and low performance due to high reject rates

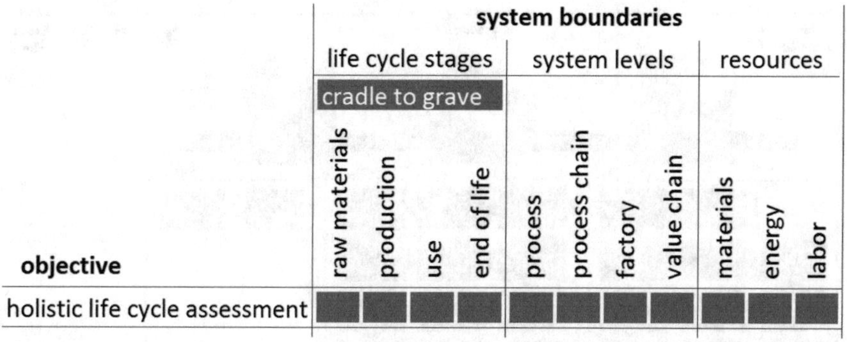

	system boundaries		
	life cycle stages	system levels	resources
	cradle to grave		
objective	raw materials	production	use	end of life	process	process chain	factory	value chain	materials	energy	labor
holistic life cycle assessment	■	■	■	■	■	■	■	■	■	■	■

Fig. 6.3 Selection of system boundaries following from the objective to conduct a holistic life cycle assessment

the production stage. Finally, all relevant resource flows are taken into account, i.e. material, energy and labor inputs as well as resulting waste and emission outputs.

From a product perspective, the analyzed value chain starts with the utilization of steel billets as main input material provided from a raw steel supplier. The value chain covers all steps from steel billet to ready-to-install piston rods, containing various processes from the main groups reshaping, separating, joining, coating and material property modification (compare Sect. 2.1.2). In total, 32 consecutive process steps are conducted in the three factories of the value chain. The distinction between process steps is mainly due to the used machinery. A detailed overview about the conducted process steps is given in Fig. 6.4. The nomenclature used for processes follows the logic *factoryID_productID_processID*.

The production starts with the transport of steel billets from a supplier to *factory 1*. This factory mainly performs reshaping and separating processes as well as thermal treatment to improve the mechanical properties of the material. The products already receive an outer shape which is close to the final product supplied to the customer of the value chain. In *factory 2* a focus is put on surface treatment processes, containing several grinding and polishing activities. As crucial functional product change, a chromium layer is applied in order to ensure a high surface durability. *Factory 3* then performs two finishing steps before the piston rods are supplied to a customer.

The holistic focus of the analysis does also necessitate to consider a broad set of performance indicators, covering all evaluation dimensions and different system levels. Thus, in general all of the PIs defined in Sect. 4.4.1 are taken into account. In order to build up a system model, data needs to be acquired as described in Sect. 4.3.2. For the companies involved it was possible to collect detailed data on all system levels, exploiting different acquisition strategies including expert interviews, document analyses and measuring and metering. Still, in some cases estimated data is used to fill existing gaps in data availability. Load profile clustering (LPC) as method to automatically interpret dynamic resource demand profiles (compare Sect. 4.3.2.2) is not applied in this case study due to the limited availability of dynamic resource demand data. Though, resource flow demands to manufacture a product are mostly

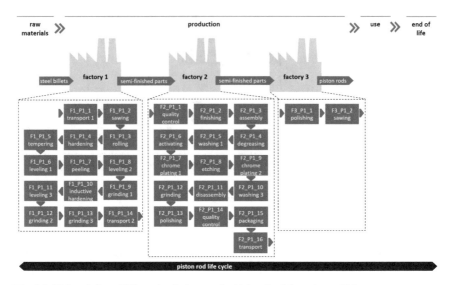

Fig. 6.4 Value chain and life cycle of piston rod with involved factories and inherent process steps

derived from documented long-term average data. Due to potential uncertainties resulting from low data quality, sensitivity analyses are conducted to identify hotspots (compare Sect. 4.4.2.2) and to ensure the validity of results. Regarding product related data, Table 6.1 summarizes the considered parameter values.

On process level, several parameters are used to characterize each process (compare Table 4.6 in Sect. 4.3.3.2). Table 6.2 shows an exemplary excerpt of data collected for process *F1_P1_7: peeling*.

In terms of resource demands, Table 6.3 summarizes the quantitative flows related with the process. Apparently, several resource inputs such as electricity and water are needed for value creation. Also, tool wear is considered here by allocating the overall demand for cutting tools to each single product. A comparison of intermediate resource input and output masses reveals a significant share of removed material. This results in a waste output flow of ~13 kg steel chips. Due to a certain share of defective parts (*QR* of 99%), ~2 kg of product are on average re-entering the process for rework.

In terms of resource related parameters (compare Table 4.4 in Sect. 4.3.3.2), company specific cost factors are used for all resources demanded along the value chain. Further, environmental impact factors from the *ecoinvent* database are multiplied with each resource flow in accordance with the type and origin of resources

Table 6.1 Product related model parameters for piston rods	Parameter	Value	Unit
	Product ID (ID_p)	1	–
	Product name (N_p)	Piston rod	–
	Product mass (m_p)	90.6	kg

Table 6.2 Process related model parameters for process *F1_P1_7: peeling*

Parameter	Value	Unit
Process ID (ID_{proc})	7	–
Process name (N_{proc})	Peeling	–
Process group (G_{proc})	Separating	–
Process type (T_{proc})	Peeling	–
Number of machines (NOM_{proc})	1	–
Batch size (BS_{proc})	1	pcs./batch
Quality rate (QR_{proc})	99	%
Rework rate (RR_{proc})	100	%
Successor for rework (SCS_{proc})	Process 7: peeling	–
Processing time (PT_{proc})	120	s

Table 6.3 Resource input-output table for producing one product at process *F1_P1_7: peeling*

Resource inputs			
w	IE_01: electricity	8.693	kWh
	IE_03: compressed air	0.284	kWh
	IM_06: coolant and lubricant	0.058	l
	IM_09: process water	0.649	l
	IM_15: cutting tools	0.001	kg
	IM_40: labor	96.970	s
Intermediate resource inputs and outputs			
z	IM_01: semi-finished product, in	210.222	kg
	OM_01: semi-finished product, out	-195.280	kg
	OM_01_RW: semi-finished product for rework, out	-1.973	kg
Waste outputs			
r	OE_09: waste heat	-8.887	kWh
	OM_07: coolant and lubricant, waste	-0.058	l
	OM_09: waste water	-0.064	l
	OM_10: evaporated water	-0.585	l
	OM_15: cutting tools, used	-0.001	kg
	OM_23: steel chips	-12.970	kg

Table 6.4 Parameter settings for end of life stage

Parameter	Value	Unit
End of life option ($OP_{p, EOL}$)	Recycling	–
Material masses in product ($mm_{p, EOL}$): steel	90.6	kg
Collection rate ($CR_{p, EOL}$)	90	%
Recycling rate ($RR_{p, EOL}$)	70	%
Quality degradation rate ($QDR_{p, EOL}$)	100	%

in order to reflect the raw material extraction stage. The product use stage is not explicitly modeled due to highly individual use scenarios depending on the system that is using the piston rod. In contrast, the end of life stage is considered in the model. It is assumed that 90% of the product can be collected for recycling, while on average 70% of the collected material can be recovered. This secondary material is assumed to replace primary raw material inputs and is therefore rewarded with credits according to the applied substitution method (compare Sect. 4.3.3.2). Table 6.4 summarizes the parameter settings that are used for end of life modeling.

With the *decision support toolbox* described in Chap. 5, a complete system model is built up using the following files:

- five connected *single factory files* as core of the toolbox, one for each of the three factories involved in the value chain plus one each for raw material extraction stage and end of life stage
- five *eco impact files*, containing impact factors for resources used in each *single factory file*
- one *value chain file*, aggregating PI results from *single factory files*
- multiple *benchmarking files* to carry out benchmarkings for specific processes
- one *KBS file* to identify case specific improvement measures
- one *MCDA file* to support the final selection of measures.

In the following, the utilization of all files is described, starting with the evaluation of the initial system performance.

6.1.2   *Phase II: Evaluation*

In the evaluation phase, a performance analysis followed by a hotspot identification are first conducted to assess the technical, economic and environmental system performance (compare Sect. 4.4.1). On the one hand, the performance of every single factory is assessed independently from each other. An example of PI results for *factory 1* is presented in Fig. 6.5. A total energy intensity of ~235 kWh is calculated for the manufacturing of one piston rod in this factory. From Fig. 6.5, the distribution of this energy demand along the factory internal process chain can be read off. A PI-based hotspot identification (compare Sect. 4.4.2.1) identifies three processes

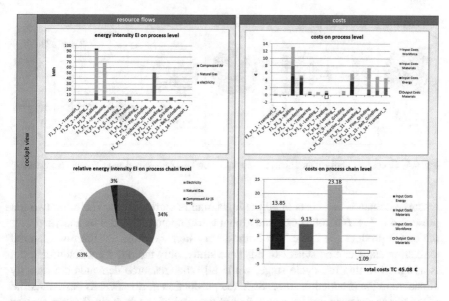

Fig. 6.5 Excerpt of performance analysis (cockpit view – energy, costs) for *factory 1* (screenshot)

as main drivers in terms of energy: process *F1_P1_3: rolling*, process *F1_P1_4: hardening* and process *F1_P1_10: inductive hardening*. Their dominant positions can be traced back to energy intensive material heat ups to several hundred degree Celsius. In terms of processes *F1_P1_3* and *F1_P1_4*, the heating is realized using natural gas combustion, whereas for process *F1_P1_10* electricity is used instead. As a consequence, the share of natural gas from the absolute energy intensity (in kWh) amounts to ~63%. From a cost perspective, additional processes move into focus, as the costs for a kWh of electricity are higher than for natural gas. In addition, other resources are taken into account as well. Processes *F1_P1_12* and *F1_P1_13*, performing grinding operations, as well as process *F1_P1_14: transport*, now also contribute with significant shares to the factory internal total costs of ~45 € per piston rod. Apparently, costs for materials and in particularly labor constitute the economic importance of these processes.

By importing the results of all *single factory files* into the *value chain file*, the performance analysis reveals the distribution of hotspots along the value chain and the entire product life cycle (compare Fig. 6.6). The calculation of technical PIs, i.e. lead time *LT*, quality rate *QR*, material utilization efficiency *MUE* and energy intensity *EI*, is only reasonable for the production stage. Concerning the overall *LT*, *factory 1* with ~60% and *factory 2* with ~40% mainly contribute, while *factory 3* has a negligible impact. A similar pattern can be observed for the distribution of energy demands. In terms of *QR*, particularly in *factory 2* a need for action is obvious due to a total *QR* of only ~70%. Regarding *MUE*, *factories 1* and *3* feature significant material losses. In particular for *factory 3* this is regarded as critical, because all material to be removed at such late stage in the value chain has already entailed

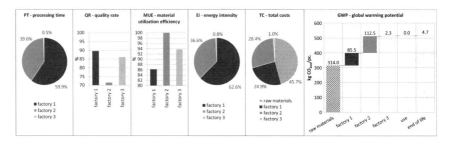

Fig. 6.6 Excerpt of performance analysis (cockpit view) for value chain and life cycle (screenshot)

resource demands and costs in all previous manufacturing steps. The other two PIs presented in Fig. 6.6, total costs and global warming potential, can be analyzed over the entire product life cycle. For both, a very high relevance of the raw material extraction stage can be stated. In this case study, only the raw steel resource input is assigned to this life cycle stage, while all other resource demands are directly allocated to the production stage. Still, the production of raw steel as basic material of the piston rod accounts for nearly half of the total life cycle costs. Further, *factory 1* and *2* contribute with ~25% each for costs and 35% each for *GWP*. In contrast, *factory 3* as well as the end of life stage are of minor importance in terms of these PIs.

Although first hotspots can be identified from the PI values and their visualization, a conduction of sensitivity analyses is regarded as meaningful due to so far unknown cause-effect relations along the studied value chain. As described in Sect. 5.2.2, *single factory files* offer a function to carry out such analyses automatically, whereby only a variation factors must be defined for each run. All changeable input parameters of the model are then automatically multiplied with this factor. Resulting PIs can be analyzed on all system levels in order to identify critical parameters and critical processes. The results for two runs using variation factors of +20 and -20% are presented in Fig. 6.7 with regard to total costs *TC* as target PI. Only model parameters of the production stage are changed, as parameters from other life cycle stages are rather outside the control of involved companies. In Fig. 6.7a, critical system parameters are shown that entail a change in *TC* of more than 8% for a 20% parameter change. Apparently, critical parameters with the highest impacts on costs relate to quality rates of single manufacturing processes. This is not surprising, as a decrease in *QR* directly results in a higher share of defective parts and thus resource wastage. Accordingly, reducing failure rates of processes is of high importance and should be granted priority in order to increase resource efficiency. Apart from quality rates, the cost factor for steel in *factory 1* is also of high relevance. Although cost factors *CF* are not directly related to the actual resource efficiency, they are usually of high economic importance. As most companies only have limited influence on these parameters through their purchase department, *CF* are not regarded as primary field of action in this concept. Beneath cost factors, resource flow quantity related parameters move into focus as well. In the first place, the demand for (raw) steel in process

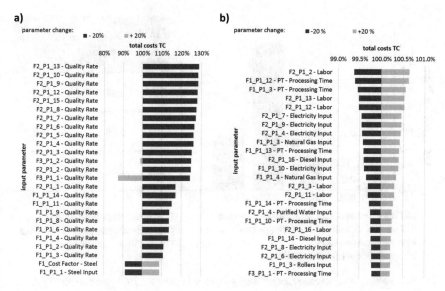

Fig. 6.7 Sensitivity analyses for piston rod value chain, initial set of most critical parameters (**a**) and filtered set excluding quality rates and cost factors (**b**) (screenshot)

F1_P1_1 is of major importance. In order to detect additional resource flows and process parameters which can be better influenced by companies, a further filtering of parameters is conducted. Figure 6.7b shows the influence of relevant parameters from *factories 1* and *2* which either relate to time or resource input quantities. For many processes, the inputs of labor and electricity as well as the processing times constitute important levers for cost reductions.

For critical parameters relating to electricity demands, a potential quantification can be carried out following the guidance provided in *step 6: potential quantification* (compare Sect. 4.4.3). Processes with electricity inputs as critical parameter are process *F2_P1_7: chrome plating 1* and *F2_P1_9: chrome plating 2* of *factory 2*. The latter is used to exemplify the benchmarking procedure against best available technology *BAT* and physical optimum *PhO*.

Figure 6.8 compares its specific energy consumption *SEC* per coated surface of 1.27 kWh/m^2*μm with the physical optimum (0.02 kWh/m^2*μm) and *BAT* values taken from different literature sources. As the *BAT* values range from 0.95 to 3.56 kWh/m^2*μm (compare Table 4.14 in Sect. 4.4.3), the lowest value is taken here as reference for performance gap calculation. The comparison shows that the *SEC* constituted by the *PhO* is several orders of magnitude lower than all available empiric values including *BAT*. Accordingly, chrome plating can in general be regarded as rather inefficient production technology in terms of energy utilization. However, taking into account the *BAT* values found in literature, the process under assessment seems to be already in a good range.

Existing performance gaps can now be quantified in order to allow for a first estimation of achievable improvements. An organizational gap *OG*, basically expressing

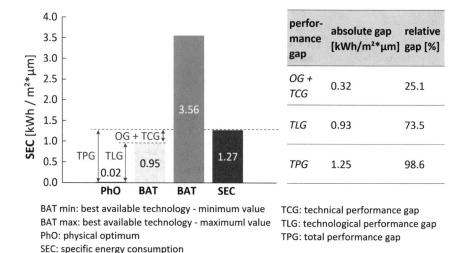

perfor-mance gap	absolute gap [kWh/m²*µm]	relative gap [%]
OG + TCG	0.32	25.1
TLG	0.93	73.5
TPG	1.25	98.6

BAT min: best available technology - minimum value   TCG: technical performance gap
BAT max: best available technology - maximuml value   TLG: technological performance gap
PhO: physical optimum   TPG: total performance gap
SEC: specific energy consumption

Fig. 6.8 Potential quantification for electricity input of process *F2_P1_9: chrome plating 2*

energy wastage caused by organizational conditions, cannot be calculated directly. This is due to a lack of information about state-based resource demand levels in terms of electricity and thus value adding energy intensities *VAEI*. However, improvement potentials due to organizational reasons are assumed, as the machine's utilization rate *MUR* is comparably low with ~60%. This does typically result in a certain share of non-value adding resource demands. In this example, *OG* is calculated together with the technical gap *TCG* as 0.32 kWh/m²*µm or 25.1%. This amount of energy could potentially be saved by using the best technology available at the market. Still, uncertainties due to different framework conditions might exist and cannot be quantified at this stage. A technological gap *TLG* of 0.93 kWh/m²*µm or 73.5% can further be calculated by comparison of *BAT* with *PhO*, i.e. this amount of energy could theoretically be saved by using an *optimal* technology. In practice, this potential is not fully exploitable due to inefficiencies going along with every manufacturing technology available today.

6.1.3   Phase III: Improvement

In *phase III: improvement*, the identification, evaluation and selection of improvement measures is carried out. Therefore, information from *single factory files* is imported to a *KBS file* in a first step (compare Sect. 5.2.6). An automatic identification of improvement measures is then conducted by means of the knowledge-based system. Figure 6.9 shows a selection of measures identified for some of the most critical parameters in terms of the PI energy intensity. These parameters relate to either

PI	critical parameter	top measures ML=1	well fitting measures 1 > ML ≥ 0.75	identified measures partly fitting measures 0.75 > ML > 0
energy intensity EI	F1_P1_4 - natural gas input	m33-4, m43-3, m44-2	m22-1-1, m22-1-2, m22-1-3	m11-1, m22-1, m24-1, m24-2, m24-3, m24-7, m31-1, m31-2, m31-3, m31-4, m41-1, m41-2, m42-1, m42-2, m42-3, m42-4, m42-5, m43-1, m43-2, m43-4, m44-5, m44-6
energy intensity EI	F1_P1_10 - electricity input	m33-4	m22-1-1, m22-1-2, m22-1-3, m43-3, m44-2	m11-1, m22-1, m24-7, m31-1, m31-2, m31-3, m31-4, m41-1, m41-2, m42-1, m42-2, m42-3, m42-4, m42-5, m43-1, m43-2, m43-4, m44-5, m44-6
energy intensity EI	F2_P1_7 - electricity input			m11-1, m22-1, m24-1, m24-2, m24-3, m24-7, m31-1, m31-2, m31-3, m31-4, m41-1, m41-2, m42-1, m42-2, m42-3, m42-4, m42-5, m43-1, m43-2, m43-4, m44-5, m44-6
energy intensity EI	F2_P1_9 - electricity input			m11-1, m22-1, m24-1, m24-2, m24-3, m24-7, m31-1, m31-2, m31-3, m31-4, m41-1, m41-2, m42-1, m42-2, m42-3, m42-4, m42-5, m43-1, m43-2, m43-4, m44-5, m44-6

Fig. 6.9 Excerpt of identified improvement measures for PI energy intensity in the *KBS file* (screenshot)

natural gas or electricity inputs of rolling, thermal treatment and chrome plating processes.

Apparently, for critical parameters related to the thermal treatment processes *F1_P1_4* and *F1_P1_10*, several top measures and well-fitting measures are identified. For the other critical parameters related to chrome plating processes *F2_P1_7* and *F2_P1_9* only partly fitting measures are found, as no specific measures for chrome plating processes are saved in the database. Table 6.5 gives a brief overview about top measures and well-fitting measures that are proposed by the system. Four of them address technological performance gaps by promoting a shift to another process technology, e.g. substitute a furnace hardening process with an inductive hardening process. Two measures, in contrast, address a technical performance gap. Measure *m33-4* aims at a usage of unused waste heat to pre-heat the product before thermal treatment. Measure *m43-3* proposes to minimize the space volume which needs to be heated up to process temperature. In some cases (e.g. measures *m22-1-1* and *m44-2*), proposed measures may also be contradictory. This is due to the fact that many measures are not generally beneficial for each manufacturing system but their advantages and drawbacks do highly depend on the specific use case as well as on the objectives pursued.

Apart from these measures, a longer list of partly fitting measures is received, which should be taken into account as well. Still, it is meaningful to consult experts in order to clarify the feasibility and potential of these measures on the one hand and discuss further possibilities on the other hand. In the presented case study, discussions with experts have provided several further alternatives, which are presented in Table 6.6. They propose rather revolutionary approaches, which are quite case specific and could therefore not be anticipated by the KBS.

Table 6.5 Description of measures identified to improve process *F1_P1_4: hardening* and process *F1_P1_10: inductive hardening*

Number	Name	Description	Performance gap
m22-1-1	Substitute electrical energy with natural gas	Substitute electricity by natural gas input to reduce energy costs and environmental impacts	Technological
m22-1-2	Replace fuel oil with natural gas	Substitute fuel oil by natural gas input to reduce energy costs and environmental impacts	Technological
m22-1-3	Replace fuel oil with wood	Substitute fuel oil by wood input to reduce energy costs and environmental impacts	Technological
m33-4	Preheat process materials and equipment	Use waste heat from other processes to preheat materials and equipment, reducing direct energy demands of thermal process	Technical
m43-3	Minimize cooling or heating space	Minimize space volume, which needs to be heated up, e.g. by installing removable plastic screens or panels or by filling cooling space with polystyrene foam blocks	Technical
m44-2	Inductive instead of furnace hardening	Use an inductive hardening process instead of electricity driven hardening furnace	Technological

Table 6.6 Excerpt of additional measures derived in discussions with company experts

Name	Description	Performance gap
Improve raw steel quality	Use raw steel with higher quality in *factory 1* as base material for piston rod in order to reduce quality related resource wastage in *factory 2*	Technical
Reduce transport distances	Reduce transport distances between all partners by either selecting local value chain partners or increasing manufacturing depth, i.e. reducing the number of partners in the value chain	Organizational, technical
Integrate highly automated chrome plating machine	Replace conventional machinery in *factory 2* by a new machine concept for chrome plating featuring a higher degree of automation	Technical

In the following, a measure evaluation is performed (compare Sect. 4.5.2) in order to quantify related improvement potentials. Apart from the initial state (scenario *S1: baseline*), three promising measures are compared:

- a *substitution of electrical energy* input in process *F1_P1_10* by natural gas input (scenario *S2: hardening*), inducing a shift from inductive hardening to furnace hardening as proposed in Table 6.5
- an *improved raw steel quality* (scenario *S3: steel quality*) as proposed in Table 6.6
- an integration of a *new chrome plating machine* (scenario *S4: automated plating*) as proposed in Table 6.6.

For this purpose, knowledge from company internal documents, literature and expert consultation is exploited. The system model is adjusted for each scenario, changing parameters and model structure where required. To give an example, the most relevant applied changes of model input parameters for scenario *S2* are displayed in Table 6.7. The overall electricity input of process *F1_P1_10: inductive hardening* is reduced to 5% compared to scenario *S1: baseline*. This amount is assumed to be needed for machine operation (controls, handling, exhaust air etc.). Instead of electricity, natural gas is now used for heating. The total heat quantity (in kWh) increases by 25% due to higher heat losses going along with furnace hardening compared to induction hardening. However, a kWh of natural gas features both economic and environmental benefits compared to a kWh pf electricity. Further, the processing time is assumed to increase from 600 to 3,600 s per batch.

Based on the parameter estimations, a first evaluation of overall impacts can be derived for each measure. Figure 6.10 shows a visual comparison of scenario *S2* towards the baseline scenario *S1* from perspective of *factory 1*.

Apparently, improvements may be achieved regarding *GWP* (-6.7%) and *TC* (-1.2%) by shifting to a furnace hardening process. As a drawback, *EI* increases by 5% and the sum of *PTs* rises by 6.4%. PIs of other factories or life cycle stages are not influenced by the measure. As a consequence, relative PI changes from a life cycle perspective are less significant (compare Fig. 6.11).

An overview about all measures considered at this point is compiled in order to receive a clear comparison of the alternatives. As displayed in Fig. 6.12, all three measures entail improvements for almost all PIs. Estimated improvements for the revolutionary measures applied in scenarios *S4: automated plating* and in particular

Table 6.7 Excerpt of adjusted model input parameters for *scenario S2: hardening*

Scenario	Name	Process	Input parameter	Original (*S1*)	Adapted (*S2*)
S2	Hardening	*F1_P1_10: inductive hardening*	Electricity input	2,196,000 kWh/a	109,800 kWh/a
			Natural gas input	0 kWh/a	2,607,750 kWh/a
			Processing time	600 s/batch	3,600 s/batch

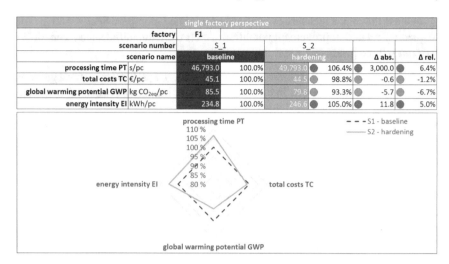

Fig. 6.10 Evaluation of scenario *S2* from perspective of *factory 1* (screenshot)

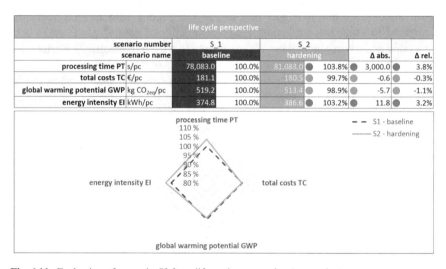

Fig. 6.11 Evaluation of scenario *S2* from life cycle perspective (screenshot)

S3: steel quality are significant compared to scenario *S2*. For the latter case, this is urged by a considerable reduction of quality related losses in *factory 2*. Technically, the higher quality steel features less particle inclusions at the material surface leading to fewer failures in later chromium plating processes. However, raw material prices are estimated to be ~20% higher. This disadvantage is taken by *factory 1*, whereas *factory 2* profits from better quality rates. Thus, a fair business model should be elaborated from which both partners benefit. Differences regarding environmental

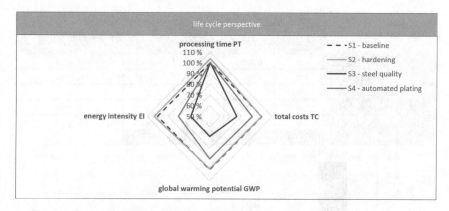

Fig. 6.12 Comparison of scenarios *S2* to *S4* with baseline scenario *S1* from life cycle perspective (screenshot)

impacts for raw steel production could not be quantified at this point but should be analyzed before measure application.

As last step of the improvement procedure, a measure selection is conducted. Due to the occurring target conflicts, a measure ranking by the PROMETHEE method is applied here to. Therefore, a manual weighting of PIs is carried out, putting highest priority to *TC* and resulting in the following PI weighting vector:

$$
\vec{w} = \begin{pmatrix} w_{PT} \\ w_{TC} \\ w_{GWP} \\ w_{EI} \end{pmatrix} = \begin{pmatrix} 0.15 \\ 0.5 \\ 0.2 \\ 0.15 \end{pmatrix}
\tag{6.1}
$$

After weighting, the consecutive steps of the PROMETHEE method are carried out according to Sects. 2.3.3 and 4.5.3.2 *Gaussian preference functions* are used here exemplarily for all PIs with a σ of 0.5. Figure 6.13 shows the results for measure ranking according to their net flows. Scenario *S3*, promoting an improved raw steel quality, is clearly identified as preferred option, followed by scenario *S4*. Scenario *S2*, in contrast, features only slight advantages against the baseline scenario *S1*. If no further criteria such as invest, implementation timeframe etc. are taken into account, an implementation of measures from left to right is recommended. Still, a detailed technical planning and validation of measure evaluation is required before real-world implementation.

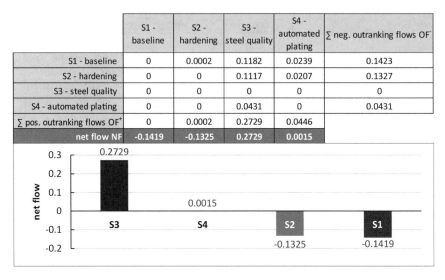

Fig. 6.13 Ranking of scenarios with PROMETHEE method according to their net flows (screenshot)

6.2 Case Study 2: Axle Journal Manufacturing

The second case study deals with the manufacturing of axle journals as parts of drive shafts for automobile power trains. A drive shaft transmits the torque of the motor from the gearbox or differential to the wheels. It must also compensate angle and length changes caused by deflections and steering movements. Figure 6.14 indicates the typical position of axle journals as connection between drive shafts and wheels in automobile power trains.

6.2.1 Phase I: Modeling

Regarding system boundaries and objectives, the following decisions are taken:

- *Axle journals* constitute the analyzed product type due to large related production volumes of several millions per year in this factory and thus a high relevance for the company.
- Main objectives of the analysis comprise a *technical assessment* of the process chain performance. In particular, hotspots related with energy utilization shall be identified.
- In terms of resource flows, only *electrical energy inputs as well as material inputs and outputs* related to the product's main material are taken into account. All minor energy flows, material flows and human labor are outside the scope for this analysis.

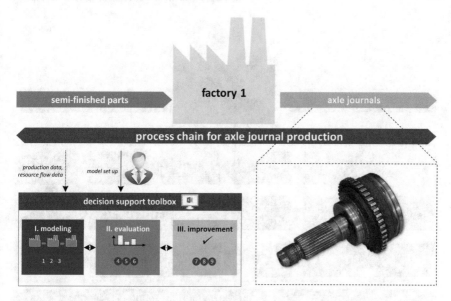

Fig. 6.14 Process chain for manufacturing of axle journals

- Concerning product life cycle stages and manufacturing system levels, the analysis is limited to the *process chain within one specific factory* of the production stage.

In this use case, an intra-factory process chain is selected as system boundary for the manufacturing of axle journals. Within the selected production line, one specific type of axle journals for medium sized automobiles is manufactured. Apart from the selected production line, additional lines with similar setup exist within the same factory. Accordingly, a certain transfer potential of the findings and thus a high relevance of the results for the whole factory is assumed. As main raw material input, the first process receives a semi-finished axle journal made of a steel alloy. 16 steps are then conducted along the process chain to receive a ready-to-install axle journal. Most machines apply separating processes (turning, drilling, washing). Further, reshaping (rolling) and material property modification (hardening) as well as inspection and handling processes are carried out. An anonymized overview about the process chain is given in Fig. 6.15.

The characterization of machines and processes is oriented towards the descriptions provided in Sect. 4.3.3.2. Data from detailed meterings of electricity demands is used for all machines within the process chain in order to achieve a high degree of transparency about energy related hot spots. Load profile clustering (LPC) (compare Sect. 4.3.2.2) is used to automatically interpret the metered load profiles. Figure 6.16 exemplifies the procedure for the given use case by means of process *F1_P1_11*. In a first step, the product is identified using the known processing time of 18 s. The identified processing interval features an average electrical load of ~9.5 kW.

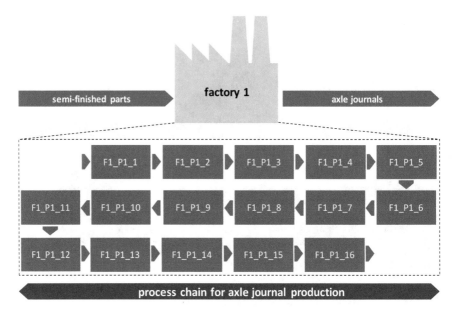

Fig. 6.15 Process chain for axle journal production with inherent process steps (anonymized)

Based on the identified interval and the underlying k-means clustering of load values, the toolbox is able to provide further information resulting from LPC application (compare Fig. 6.17). State-based resource demand levels for both value adding state (processing) and non-value adding state (waiting) are calculated and can be used

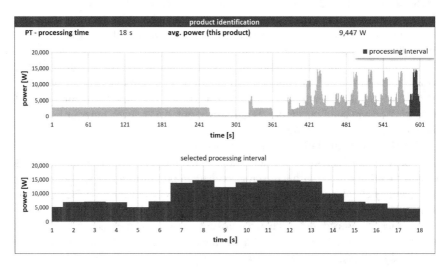

Fig. 6.16 Automatic identification of processing interval in electrical load profile of process *F1_P1_11* (screenshot)

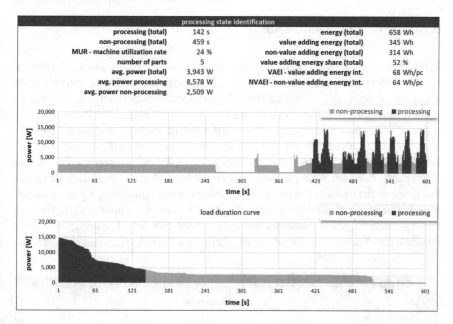

Fig. 6.17 Automatic identification of further processing intervals and derivation of state-based resource demand levels in electrical load profile of process *F1_P1_11* (screenshot)

as input for system modeling. Further, first estimations in terms of value adding and non-value adding energy intensities (*VAEI, NVAEI*) are provided automatically, indicating a value adding energy share of only ~52%. Besides, a machine utilization rate *MUR* of only 24% during the analyzed time period is derived and visualized using a load duration curve. Such curves can be received by sorting load values from highest to lowest. They ease the visual interpretation of energy meterings and can, especially on higher system levels, also be used to identify improvement measures related to load management (Herrmann et al. 2013). In contrast to resource demand levels, calculated PI values for *VAEI, NVAEI* and *MUR* are only valid for the considered time period, which is ten minutes in this example and might thus not be representative.

With the *decision support toolbox*, a complete system model is built up using the following files:

- one *single factory file* representing the analyzed factory
- one *benchmarking file* to carry out a benchmarking for specific processes

Input-output models characterize each process with a focus on main material and electrical energy flows, also using the results from LPC application.

6.2.2   Phase II: Evaluation

In the evaluation phase, a performance analysis is conducted to assess the technical system performance (compare Sect. 4.4.1). In accordance with the rather narrow scope of the analysis, mainly PI from the technical dimension are considered on process and process chain level. Figure 6.18 shows an excerpt of PI dealing with capacity utilization. Accordingly, the major share of the total lead time *LT* can be traced back to process *F1_P1_8*. However, the analysis of gross cycle times *GCT* reveals potential bottlenecks at processes *F1_P1_2*, *F1_P1_15* and *F1_P1_16*. The discrepancy between both PI can be explained by different batch sizes of related machines on the one hand and differing numbers of machines working in parallel on the other hand. As an example, process *F1_P1_8* features a batch size of >1,000 parts, while process *F1_P1_3* features a batch size of one part only. However, the latter is conducted by ten machines working in parallel. As the utilization rates of machines

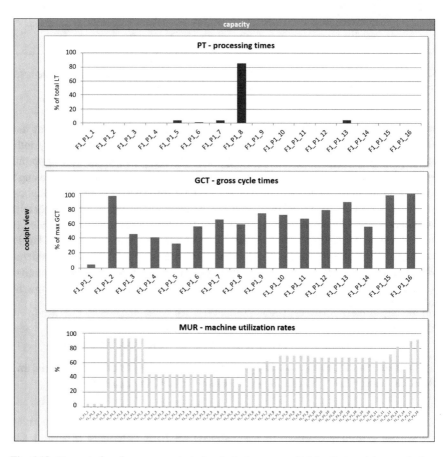

Fig. 6.18 Excerpt of performance analysis (cockpit view – capacity) for *factory 1* (screenshot)

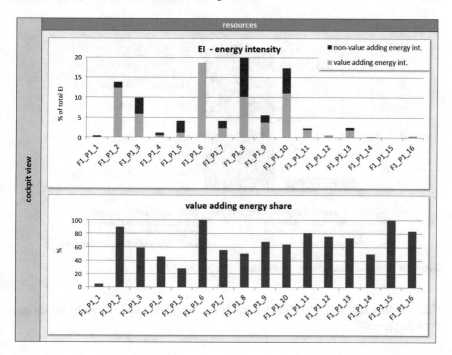

Fig. 6.19 Excerpt of performance analysis (cockpit view – energy) for *factory 1* (screenshot)

basically correspond to their *GCT*, highly differing utilization rates between ~5 and ~93% can be stated, whereas the average utilization is calculated to ~59%.

As mentioned in Sect. 6.2.1, one major objective is the identification of hotspots related with energy demands. Figure 6.19 provides an overview about *VAEI* and *NVAEI* as well as relative shares of value adding energy for each process. Altogether, a value adding share of ~70% is calculated for the entire process chain. The highest absolute energy losses can be stated for processes *F1_P1_8* and *F1_P1_10*. From a relative perspective, high saving potentials can also be assumed for some other processes.

Due to the limited complexity of the assessed process chain, a PI based identification of hotspots is regarded as sufficient in a first step (compare Sect. 4.4.2.1). Accordingly, processes with the highest direct electricity inputs are considered as critical processes. In order to judge about achievable improvements, these processes are benchmarked against *BAT* and *PhO* values. In Fig. 6.20, the benchmarking for process *F1_P1_8* against *PhO* is exemplarily presented, revealing a total performance gap *TPG* of ~91%. This theoretical saving potential can be broken down into an organizational gap *OG* of ~50%, which can be achieved by organizational measures such as a better machine utilization or a machine shut down in non-production times. Additional ~41% of saving potential is due to technical and technological reasons, while a differentiation could not be conducted due to a lack of valid *BAT* values.

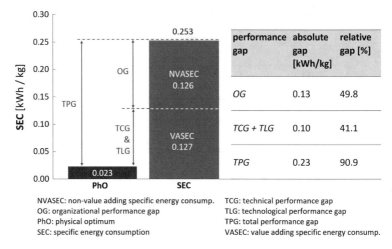

NVASEC: non-value adding specific energy consump.   TCG: technical performance gap
OG: organizational performance gap   TLG: technological performance gap
PhO: physical optimum   TPG: total performance gap
SEC: specific energy consumption   VASEC: value adding specific energy consump.

Fig. 6.20 Potential quantification for electricity of process *F1_P1_8*

Still, the benchmarking provides a first impression about achievable energy savings for this process.

6.2.3   Phase III: Improvement

The identification, evaluation and selection of improvement measures in the third phase was conducted together with factory internal experts. Though, the knowledge-based system as introduced in Sect. 4.5.1 is not used in this case study. In Table 6.8,

Table 6.8 Excerpt of measures identified in discussions with company internal experts

Name	Description	Scenario
Improve machine utilization	Utilization rates of machines highly differ, leading to high non-value adding energy demands (~30% in total). Some machines in processes with low utilization rates can be switched off and only ramped up when needed.	*S2*
Insulate machine	Heat losses in process *F1_P1_8* can be reduced by installing a better insulation at the machine surface.	*S3*
Shut down machines in break times	Machines are usually kept in waiting mode during break times. As such break times amount for three hours per day, machines that are technically suited can be automatically shut down during breaks.	*S4*

a selection of promising improvement measures is presented that are evaluated as alternative scenarios.

A measure evaluation is performed (compare Sect. 4.5.2) in order to quantify the improvement potential for all measures, i.e. their impacts on the overall system performance compared to the baseline scenario *S1*. Parameter changes and structural changes are applied for each scenario using knowledge of the process owners. As an example, the total number of used machines is reduced by ~25% in scenario *S2*, resulting in an increase of the average machine utilization rate from ~59 to ~78%. This induces a reduction of the *EI* per product of ~8%. Such dramatic reduction of used machines is possible due to existing over capacities in the analyzed process chain. In several process steps (e.g. process *F1_P1_3*), excess machines are operated in order to minimize the risk of throughput reduction due to machine failures. However, applied assumptions are very conservative and changes in some processes in recent years have changed processing times. The operation strategy for the machinery has, in contrast, not been adapted regularly. Still, the optimal number of machines to keep as backup can only be found by dynamic material flows simulation (compare Sect. 2.4.6) or testing in daily practice. The resulting improvements in terms of PI changes for all measures are compared in Fig. 6.21. As the evaluated scenarios focus on the improvement of energy utilization, only improvements in this category are expected. The greatest effect is obtained by a machine shut down in break times in scenario *S4*, reducing the *EI* by ~9.9%. In comparison, the application of additional insulation for the machine used in process *F1_P1_8* does only have a slight effect of ~2% improvement. However, before an implementation of one or several of the presented measures can be planned, further cost evaluations are highly recommended to judge about the economic benefits of each measure.

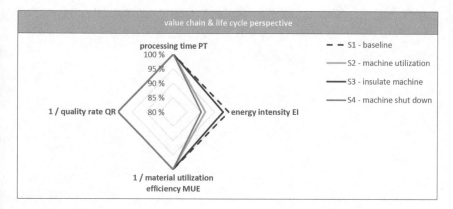

Fig. 6.21 Comparison of scenarios *S2* to *S4* with baseline scenario *S1* (screenshot)

References

Herrmann C, Posselt G, Thiede S (2013) Energie- und Hilfsstoffoptimierte Produktion. Springer Vieweg, Berlin, Heidelberg

Liebherr Hydraulikbagger GmbH A 916 Compact Litronic. https://www.liebherr.com/de/deu/produkte/baumaschinen/erdbewegung/mobilbagger/details/304405.html#lightbox. Accessed 10 Aug 2019

Chapter 7
Summary and Outlook

In this chapter, a summary of the presented concept is first provided in Sect. 7.1. It is then evaluated in Sect. 7.2 with focus on criteria applied to state of the art approaches and on identified research gaps. Finally, an outlook on possible future research is given in Sect. 7.3.

7.1 Summary

The developed concept provides decision support to increase resource efficiency in manufacturing value chains. Against the background of growing global resource demands and environmental pollution, it intends to contribute towards a sustainable development. Thus, a holistic concept to analyze and improve industrial resource efficiency is proposed, taking into account the complexity of today's global value chains and supply networks. In this context, two research questions are raised, asking for a combination and extension of existing methods with the objective to provide decision support, i.e. to allow for better manufacturing related decisions while reducing efforts for decision making. Accordingly, in Chap. 2 the need for a more resource efficient manufacturing is argued. Several well-established methods and tools for manufacturing system modeling, evaluation and improvement are presented, which set different scopes and priorities. From the state of the art, the author concludes that a support for decision makers is required to predict the local and global consequences of manufacturing decisions, e.g. by means of virtual system models and decision support systems. In Chap. 3, existing approaches to provide decision support for a resource efficient manufacturing are presented. Their comparison reveals a high maturity of available work concerning the holistic modeling and evaluation of manufacturing systems on different system levels. However, actor spanning activities on value chain level and potential interdependencies over entire product life cycles are often neglected. Moreover, decision makers are usually not sufficiently supported in the crucial step of system improvement under consideration of competing technical, economic and environmental targets. As result, a need to create holistic system models and provide decision support in system modeling, evaluation and improvement of manufacturing systems is stated. In Chap. 4, suitable requirements for a new concept to address this research demand are formulated and translated into a continuous improvement procedure, comprising nine consecutive steps. It builds upon a combination of material and energy flow analysis, values stream mapping, life cycle costing and environmental life cycle assessment as complementary methods. The concept application is conducted using a common data basis and thus allows to derive a consistent system of performance indicators, which covers different system levels and evaluation dimensions. Entirely new methods such as load profile clustering are developed and integrated into the overall procedure in order to ease decision making while reducing the need for human expert involvement. A knowledge-based system allows to automatically identify improvement measures for a given decision situation. Their impacts can be tested by means of the developed system models. In order to facilitate concept application, a *decision support toolbox* is presented. Its implementation into a stand-alone software tool featuring a high degree of automation is described in Chap. 5. During the development, a special focus was put on a user-friendly workflow, allowing a basic tool application also for non-experts. Due to the modular toolbox architecture, its application can be tailored to individual needs determined by a specific use case. Finally, the concept application is demonstrated by means of two different use cases from the metal mechanic industry in Chap. 6. The adaptability of the concept is demonstrated by covering a broad and holistic scope in the first use case (piston rod manufacturing), while focusing on energy analyses within a single factory in the second use case (axle journal manufacturing). For both examples, significant resource saving potentials are identified with the *decision support toolbox*.

7.2   Concept Evaluation

The presented concept provides a progress beyond state of the art in terms of decision support for modeling, evaluation and improvement of manufacturing value chains. In order to put the presented concept into context of existing work, the same evaluation criteria are applied. The results and corresponding explanations for each criterion are displayed in Table 7.1, indicating a total fulfillment score of 0.85. The broad and holistic modeling approach adequately reflects the complexity of interlinked manufacturing systems with their inherent levels from single processes up to entire product life cycles from cradle to grave. Here, the main achievement relates to a seamless and synergetic coupling of existing methods, i.e. MEFA, VSM, LCC, and LCA. Due to the joint data basis, consistent results can be expected from method application. However, the underlying methods have a rather static character and

Table 7.1 Evaluation of own concept

Modeling			
Planning horizons		Management tasks on tactical and strategical level can be covered	◑
Life cycle stages		All product life cycle stages from raw material extraction until end of life can be integrated, while a focus is put on the production stage	●
Manufacturing system levels		Bottom-up modeling with processes as basic elements, can be extended up to interrelated factories on value chain level	●
Resource flows	Materials	Detailed according to MEFA and LCA methodologies	●
	Energies	Detailed according to MEFA and LCA methodologies	●
	Labor	Detailed consideration in analogy to physical resource flows	●
Decision support			
Evaluation dimensions	Technical	Broad range of technical parameters and performance indicators derived from VSM and MEFA methods such as quality rate, lead time, material utilization, energy intensity	●
	Economic	Detailed economic evaluation or resource flows oriented towards LCC method	●
	Environmental	Thorough environmental evaluation in the sense of LCA methodology basing on MEFA model and LCIA data from LCA databases	●
	Social	Not considered	○
Reference values		Benchmarking against *BAT* and *PhO* values on process level and calculation of performance gaps to give indications about improvements potentials	●
Identification of improvement measures		Automated identification without expert involvement by means of KBS, providing a broad range of specific and generally applicable improvement measures	●
Ranking of improvement measures		Integration of MADM methods to allow for decision making under consideration of competing targets and target conflicts	●

(continued)

Table 7.1 (continued)

Application

Application procedure	Detailed step by step guidance provided both for general improvement procedure and toolbox application	●
Maturity of software implementation	Decision support toolbox available as advanced prototype, application on different use cases (e.g. in research project *MEMAN*) successfully conducted	◑
Accessibility	Toolbox is in general open for non experts, no specific expert software is needed due to implementation in *MS Excel*	◑
Effort of implementation	High effort for initial data acquisition due to holistic approach, moderate modeling effort due to pre-defined model elements, rather low effort for application due to high degree of decision support and automation of toolbox	◐
Average		0.85

are not able to fully analyze the dynamic time dependent behavior of manufacturing systems. As a consequence, the concept is tailored to decision making from tactical to strategic time horizons and features limitations in terms of real-time and operational decision making and decisions based on time-dependent variables. A significant progress compared with state of the art approaches can be stated regarding decision support. A holistic assessment of resource flows from a technical, economic and environmental perspective is achieved based on underlying methods and a consistent performance indicator system. Extended value stream maps are provided to summarize performance assessments from process to value chain level. Beneath a detailed performance analysis, users of the *decision support toolbox* are actively supported in improving the system performance. Procedures to systematically identify hotspots by automated sensitivity analyses and benchmarking of processes against best available technology and physical optimal processes are introduced. They allow for a quantification of performance gaps that indicate improvement potentials. An additional progress is achieved by the development of a knowledge-based system to conduct automated and situation specific measure identifications and prioritizations. A decision support towards measure implementation is provided by an integration of multi-attribute decision making methods to detect the most promising measures and measure bundles. As a shortcoming, decision support in terms of social aspects is not given so far. Regarding concept application, the developed prototype of a *decision support toolbox* already features a high level of maturity. This aspect is underlined by its successful application within the European research project *MEMAN*. Due to its user-friendliness, the toolbox successfully lowers the barriers for an application in industrial practice. Efforts for modeling and continuous application of the concept by means of the toolbox are regarded to be comparably low. As an example, the load profile clustering method eases metering data interpretation. Further, pre-defined process models are available, avoiding time-consuming modeling from scratch. Still, a high data collection effort for initial modeling must be stated owing to the holistic approach followed. In summary, a progress regarding the establishment of continuous improvement processes for resource efficient manufacturing value chains can be stated.

7.3 Outlook

In order to increase its usefulness of the presented concept, several adaptions and extensions can be taken into account which are briefly discussed in the following.

- **Application in further industrial sectors**: The concept is tailored to the needs of discrete manufacturing with a focus on the metal mechanic sector as demonstrated by means of two case studies. The underlying methods, however, are also applicable to other industries using different production processes. Further developments could extend the pre-defined model galleries by corresponding processes, e.g. by processes used in electrical industry (consumer electronics, electrical energy storages etc.). A transfer to process industries (chemistry, pharmaceutics, food, paper, glass, steel, cement etc.) may in contrast require some methodological adaptions due to the continuous character of utilized processes. However, such development is considered reasonable owing to the high relevance of these industries in terms of resource demands (Prognos AG 2009).
- **Extension by social evaluation dimension**: As stated in Sect. 7.2, social aspects are currently not taken into account. Considering the definition of a sustainable development, a balancing between economy, society and environment should be aspired. The state of research, however, revealed that detailed social assessments are barely conducted so far. Reasons for this observation can be seen in the limited availability and general validity of data. As remarkable progresses have been achieved in the field of social life cycle assessment in recent years, an integration into the presented concept could for instance be realized by applying the UNEP/SETAC *Guidelines for Social Life Cycle Assessment of Products* (United Nations Environment Programme 2009) or by using data from the *social hotspot database* (Social Hotspot Database 2019).
- **Coupling with manufacturing system simulation**: The selected combination of modeling and evaluation methods is capable to provide holistic virtual system models but with a rather static character. An extension with material or energy flow simulation methods (compare Sect. 2.4.6) could broaden the applicability to

a greater number of short-term manufacturing decisions. Simulation models could cover different elements and levels of manufacturing systems, while a coupling of different models significantly increases the complexity (Schönemann 2017). Several approaches presented in Sect. 3.3 already beneficially combine simulation with other methods, e.g. DES with LCA. For the presented concept, an extension by simulation models would likely go along with a growing need to collect data and higher efforts for system modeling. Furthermore, the simulation as such would necessitate the utilization of additional expert software. Interfaces between the *decision support toolbox* realized in *MS Excel* and the simulation environment would need to be established. As benefits, additional performance indicator could be integrated into the presented performance indicator system, into result visualizations like extended value stream maps as well as into the knowledge-based system. This could further improve the quality of recommendations resulting from concept application.

- **Implementation into a commercial software**: The provided prototypical implementation already features a high degree of maturity and can be used to analyze and improve real-world manufacturing systems. Nevertheless, a professional implementation as commercial software tool could further improve usability and reduce bugs. A professional customer service would be needed and continuous software updates would be expected from paying customers. Such development would require a significant budget that could be granted by public funding, potential industrial customers or software companies that are interested in commercializing the underlying concept.

- **Integration into Industry 4.0 solutions**: Industry 4.0 is expected to offer various new opportunities to achieve a more sustainable manufacturing (Stock and Seliger 2016). Thus, an integration of the developed concept into cyber-physical (production) systems could be aspired (Lee et al. 2015). By building up digital twins, which are coupled to the physical world by continuous data flows, a real-time decision support to achieve a higher degree of resource efficiency could be achieved. By implementing direct control loops, promising improvement measures on different system levels (e.g. adapted processing parameters) could instantly be implemented in the physical system without further human intervention. However, a digital twin of an entire manufacturing system, value chain or product life cycle would demand for extensive metering infrastructure, which is typically not yet available in practice. Consequently, an implementation could start with the embedding of single models for selected system elements with high relevance in terms of technical, economic or environmental aspects.

- **Derivation of suitable business concepts**: The presented concept and toolbox are designed to facilitate multi-criteria analyses. If target conflicts arise between different actors on value chain level or along the product life cycle, there may be a need to fairly distribute benefits and drawbacks among involved actors (compare consequences of scenario *S3* in Sect. 6.1.3). Otherwise, actors that do not expect advantages will be hardly motivated to implement changes in their manufacturing systems. Further, risks resulting from potential changes and invests should also be distributed as soon as multiple actors benefit. Consequently, support in business model development could be provided, comprising methodological guidance and best practice examples from successfully established cooperations building on a high degree of trust between partners, e.g. in cooperative supply chain management or in eco-industrial parks (Chertow 2000).

References

Chertow MR (2000) Industrial symbiosis: literature and taxonomy. Annu Rev Energy Environ 25(1):313–337

Lee J, Bagheri B, Kao H-A (2015) A cyber-physical systems architecture for Industry 4.0-based manufacturing systems. Manuf Lett 3:18–23. https://doi.org/10.1016/j.mfglet.2014.12.001

Prognos AG (2009) Energieeffizienz in der Industrie. VDMA Forum Energie, Frankfurt am Main

Schönemann M (2017) Multiscale simulation approach for battery production systems. Springer International Publishing, Cham

Social Hotspot Database (2019) Social hotspot database. https://www.socialhotspot.org/. Accessed 11 Aug 2019

Stock T, Seliger G (2016) Opportunities of sustainable manufacturing in Industry 4.0. Procedia CIRP 40:536–541. https://doi.org/10.1016/j.procir.2016.01.129

United Nations Environment Programme (2009) Guidelines for social life cycle assessment of products. Nairobi

Printed in the United States
by Baker & Taylor Publisher Services